超實用！職場英語聽說讀寫實戰教學

上班族不能不會的 E-mail+電話英語

E-mail and Telephone Tips for Daily Life and Work

上班族不能不會的

E-mail+電話英語

E-mail and Telephone Tips for Daily Life and Work

發 行 人　鄭俊琪

總 編 輯　王琳詔

責任編輯　鄭瑜伽

英文編輯　Helen Yeh

特約作者　Timothy Daniel Ostrander・Ted Pigott

英文錄音　Helen Yeh・Doug Nienhuis

藝術總監　李尚竹

美術編輯　鄭恩如 ・ 李海瑄 ・ 周秀圓

封面設計　王瑄晴

技術總監　李志純

程式設計　李志純・郭曉琪

介面設計　陳淑珍

光碟製作　翁子雲

出版發行　希伯崙股份有限公司

105 台北市松山區八德路 3 段 32 號 12 樓

劃撥：1939-5400

電話：(02)2578-7838　傳真：(02)2578-5800

電子郵件：Service@LiveABC.com

法律顧問　朋博法律事務所

印　　刷　禹利電子分色有限公司

出版日期　民國 100 年 12 月　初版一刷

推廣特價　書 + 互動光碟：399 元

超實用！職場英語聽說讀寫實戰教學

上班族不能不會的

E-mail+電話英語

E-mail and Telephone Tips for Daily Life and Work

英語數位學習第一品牌

目錄 Contents

主編的話 From the Editors

在全球化的經濟市場中，企業經常需要與世界各國的公司進行生意往來，企業員工自然也常會遇到接待外國客戶、海外出差，及處理國際貿易等事宜，因此英語成為許多上班族不能不會的能力之一。然而，忙碌的上班族並沒有太多自修時間，花錢上補習班充實英文實力往往又難以立刻見效，常常弄得吃力不討好，半途而廢的結果自然就不在話下。因此，本書針對所有上班族最需要的兩項英語溝通技能——e-mail 和電話——來設計，將職場環境最常遇到的情境收錄其中，讓上班族能有效率掌握這兩種最重要的溝通技巧。

首先，在「停看聽篇」中除了告訴你應對往來的基本禮儀外，還介紹了基本 e-mail 寫作重點與電話英語，在針對不同情境主題學習之前，你可以在這裡找到任何情況皆可能用到的實用句。在「英文 e-mail 不可不知的祕訣」中，我們整理了商業書信往來中最常見的句型，依照「開場」、「正文」、「結尾」分類，你能迅速查到可參考的實用例句。而在「電話英語不可不知的祕訣」中，我們針對電話現場的各種情況，從接通到結束，整理出最常用的應對英語，讓你快速掌握各種電話情境。

接著，我們就職場中，實際上最常用到英語的情況，分成「人際關係篇」、「商務往來篇」、「貿易活動篇」、「售後處理篇」四個部分，共三十個情境主題。每個主題當中皆包括 e-mail 的往返和電話英語的應對技巧。以 Unit 16 詢問產品為例來說明我們的規劃及想法：

Ⅰ. **詢問產品一定要會的單字片語：**

收錄與主題相關且必學的重要單字片語，這些字詞可應用在之後的 e-mail 或電話中。

Ⅱ. **詢問產品一定要會的句型：**

針對每個主題的情境來設計句型，比如在「詢問產品」中，通常會提到如何得知聯絡資訊、表達對產品的興趣、可能的顧慮，以及索取相關資料等，我們就介紹表達這四種情況最實用的句型，並且應用到之後的 e-mail 與電話英語中。

Ⅲ. 如何用 e-mail 來詢問產品及回覆產品詢問：

延續前面一定要會的句型，我們將重點整合在 e-mail 當中，除了一篇實用的範文外，並收錄一篇針對此範文所做的回應。我們在不同主題的範文中，試圖包括各種不同的情境，以應付上班族各種不同的需求。

Ⅳ. 用電話詢問產品一定要會這樣說：

同樣地，我們針對此主題推演出各種可能的情況，並且提供適當的問答，你能從小標題中快速找到各種實用的電話英語來做靈活運用。

Ⅴ. 如何用電話來詢問產品：

在這篇對話當中，我們設計了各種不同的情況，包括與個別主題相關的內容，還有一些實用的電話英語。除了閱讀書上的內容，我們建議你也可以透過 MP3 或互動光碟來加強聽力與口說的能力，做好隨時應對的準備。

Ⅵ. 換你試試看：

在 e-mail 與電話英語之後，我們都設計了「換你試試看！」的小測驗，前者是訓練你的寫作能力，後者除了寫作的句法觀念外，同時還加強聽力與口說的能力，建議你可根據我們的設計，先依提示寫出完整的句子，再聽 MP3 所播放的音檔加強聽力，最後可以跟著外籍老師的錄音來做互動，當聲音播放 A 的時候，可以想想 B 要如何回答，同時訓練聽力與口說能力。

上班族平時工作已經相當辛苦，在努力工作力爭上游之餘，希望我們這本《上班族不能不會的 E-mail + 電話英語》可以讓你更輕鬆地加強英語溝通能力，提升職場競爭力。

如何使用光碟 User's Guide

系統建議需求

[硬體]

* 處理器 Pentium 4 以上（或相容 PC 個人電腦之處理器 AMD、Celeron）
* 512 MB 記憶體
* 全彩顯示卡 800*600 dpi（16K 色以上）
* 硬碟需求空間 200 MB
* 16 倍速光碟機以上
* 音效卡、喇叭及麥克風（內建或外接）

[軟體]

* Microsoft XP、VISTA、Win 7 繁體中文版系統
* Microsoft Windows Media Player 9
* Adobe Flash Player 10

光碟安裝程序

1. 進入中文視窗，將光碟片放進光碟機。
2. 本產品備有 Auto Run 執行功能，如果您的電腦支援 Auto Run 光碟程式自動執行規格，則將自動顯現【上班族不能不會的 E-mail + 電話英語】之安裝畫面。

3. 若您電腦已安裝過本公司產品，如 CNN、Live、ABC、ALL+、Biz 互動英語雜誌及各類叢書之互動光碟，則可直接點選「快速安裝」圖示進行安裝；反之，若無法選取「快速安裝」圖示，請點選「安裝」圖示，進行完整安裝。
4. 當您要移除本光碟請點選「開始」，選擇「設定」，選擇「控制台」，選擇「新增／移除程式」，並於清單中點選「上班族不能不會的 E-mail + 電話英語」，並執行「新增／移除」功能即可；若您先前是採用快速安裝模式，請於 Auto Run 畫面點選「解除快速安裝」圖示即可移除。
5. 若您電腦無法支援 Auto Run 光碟程式自動執行規格，請打開 Windows 檔案總管，點選光碟機代號，並執行光碟根目錄的 autorun.exe 程式。
6. 如果執行 autorun.exe 尚無法安裝本光碟，請進入本光碟的 setup 資料夾，並執行 setup.exe 檔案，即可進行安裝程式。
7. 當語音辨識系統或錄音功能失去作用，請檢查音效卡驅動程式是否正常，並確認硬碟空間是否足夠且 Windows 錄音程式可以作用。
8. 麥克風設定請參照光碟主畫面的「操作及語音辨識安裝說明」中的語音辨識設定。

操作說明

點選「執行」即開啟本光碟的教學功能。說明如下：

主畫面

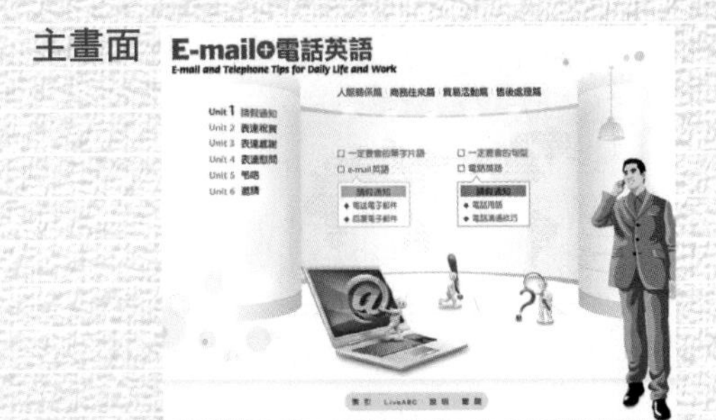

主畫面有課程單元、LiveABC 網站、索引、操作及語音辨識安裝說明及離開等圖示，點選後進入該單元。

圖解

1. 由課程目錄「一定要會的單字片語」選單進入圖解。
2. 當游標移到所教單字的圖示上，點選後會出現該單字的發音。
3. 點選畫面左下角的 Tools 圖示可叫出工具列。

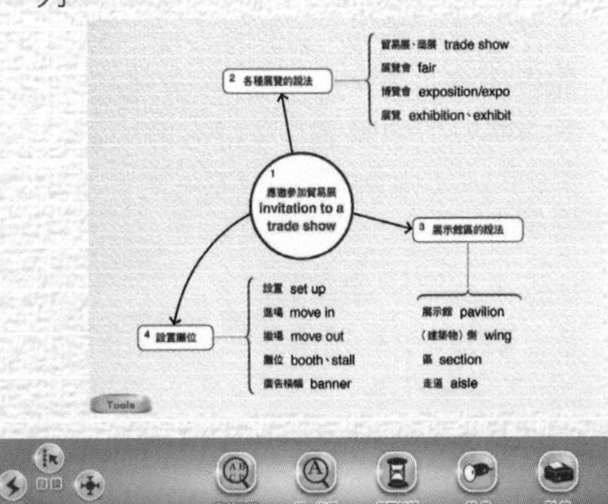

全部出現　逐一出現　反覆朗讀　錄音　列印　說明

工具列說明：

全部出現

點選後會以表單方式列出所有單字。每個單字都有音標及中譯，點選後會發音。點選下方的 Read 會自動朗讀所有的單字。若點選 Back 則會返回原畫面。

逐一出現

點選後，所有單字會伴隨發音逐一出現。一移動滑鼠則結束此功能。

反覆朗讀

點選後再點選任一單字，會反覆播放該單字的音；若想恢復一般狀態，只要再次點選本圖示即可。

錄音

點選後可打開錄音功能視窗。

錄音步驟如下：

1. 點選要進行錄音的單字，並選擇是否要在錄音前播放原音。
2. 點選「錄音」鍵。
3. 請在電腦「播放原音」後，對著麥克風唸出所選取的單字。
4. 完成錄音後，按鍵盤上的「空白鍵」結束錄音。
5. 點選「播放」鍵，即可聽到所錄的聲音。

語音辨識步驟如下：

1. 選擇要練習發音的單字，以及是否要在語音辨識前播放原音。
2. 點選「語音辨識」圖示。
3. 畫面出現「請錄音」時，對著麥克風唸出該字，若發音正確，則繼續進行下一句；若發音不準確，則會出現一視窗，您可選擇「再試一次」、「略過」或「唸給我聽」來完成語音辨識。
4. 若要結束錄音或語音辨識，請點選「停止」圖示。

説明：進入「圖解」的輔助説明頁。

一定要會的句型

點選目錄選單上藍色字體例句，即可聽發音。

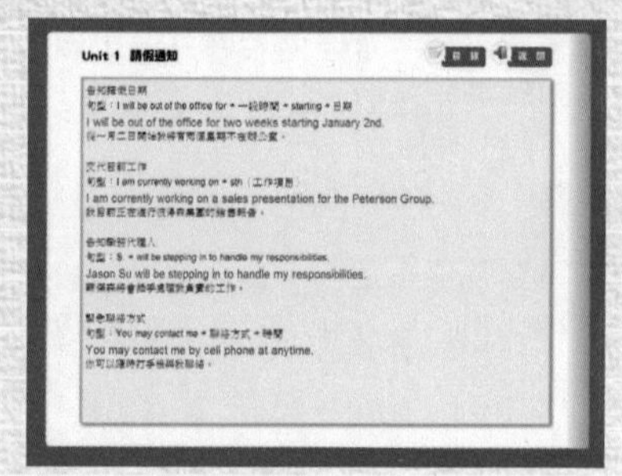

E-mail 及電話溝通技巧

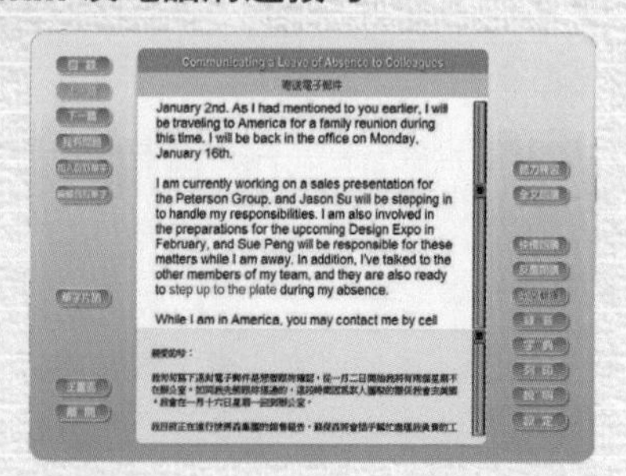

工具列説明：

全文朗讀

點選「全文朗讀」圖示，電腦將自動朗讀信件內容，若您要中途停止播放，請再點選一次本圖示或在任意處點一下即可。

文化補給站

點選後可進入該單元文化補充內容。

換你試試看

點選後可進入測驗。

角色扮演

1. 點選「角色扮演」圖示，將於圖示左側出現該場景之角色人名。請點選您所欲扮演的角色，程式將關閉該角色的聲音，由您和電腦進行對話練習。
2. 若發音不正確，會出現可以選擇「再試一次」、「略過」、或「唸給我聽」的對話框，也可調整語音辨識的靈敏度。
3. 若想中途停止角色扮演功能，請於介面任一處點選一下即可。

快慢朗讀

當您覺得對話速度太快，可以點選「快慢朗讀」圖示，再點選「全文朗讀」圖示或任一句子，朗讀速度將變慢，讓您聽得更清楚。若您開始覺得速度太慢，想恢復為一般速度，只要再次點選「快慢朗讀」圖示，取消功能即可。

反覆朗讀

1. 點選「反覆朗讀」圖示後，再選取任一句，電腦將反覆朗讀該句子，在畫面上任一處按滑鼠左鍵一下即可停止此功能。
2. 在反覆朗讀圖示上按右鍵，將出現一反覆朗讀控制視窗，您可設定句子反覆間的間隔秒數及反覆朗讀的次數。

中文翻譯

點選「中文翻譯」圖示後，將於下方出現之中文翻譯。當您點選中文翻譯框中的某句中文，則會朗讀相對應的英文句子；同樣的，點選內文中的任一句子，也會朗讀該句英文，並標示出其中文翻譯。

錄音

點選「錄音」圖示後，開啟錄音功能控制列。點選段落前方的的方框（☐）即可勾選要進行錄音的句子，然後進行錄音或語音辨識。

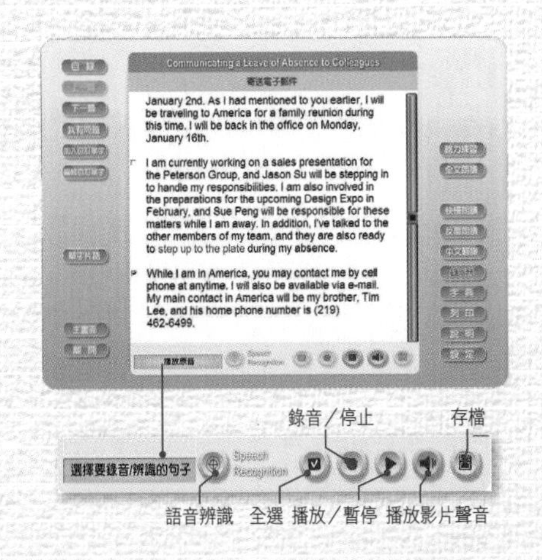

字典

當您選取「查閱字典」圖示後，在畫面下方將出現字典框，點選對話中的任一單字，字典框內會出現該單字的音標及中文翻譯，並唸出該字發音。

列印

當您選取「列印」圖示後，畫面下方將出現列印控制鍵。您可選擇「全部列印」或「局部列印」；列印內容可選擇是否包括中文翻譯。此外，本光碟還提供儲存功能，您可以選擇全部儲存或局部儲存；並選擇是否儲存中文翻譯。

說明

當您選取「說明」圖示，將開啟輔助說明頁。您可藉此瞭解本光碟的各項操作說明及用法。

學習重點

點選對話中藍色字體的學習重點，畫面下方會出現說明框，並有發音；若在開啟「中文翻譯」功能時點選，則朗讀您點選的句子。

段落朗讀

當您點選課文中的人名，程式將自動朗讀此人該段會話。若您是處於「慢速朗讀」模式，則播放該段會話時，聲音及文字反白將以小段方式出現。

我有問題

當您對文中的句子有疑問時，請點選畫面左方的「我有問題」圖示，再點選您有疑問的句子，則本程式會自動開啟 Outlook Express 系統，請將您的問題描述於句子下方。我們在收到您的來信後，將有專人為您解答。

加入及編輯自訂單字

點選加入自訂單字後，您可點選您要記錄的單字。在此，您可以進行單字學習也可以移除、列印或存取任一單字。

單字解說

在此將列出本場景之重要單字、音標、註釋、與例句，點選藍字部分會發聲。

索引

在主畫面點選本圖示，進入此畫面。

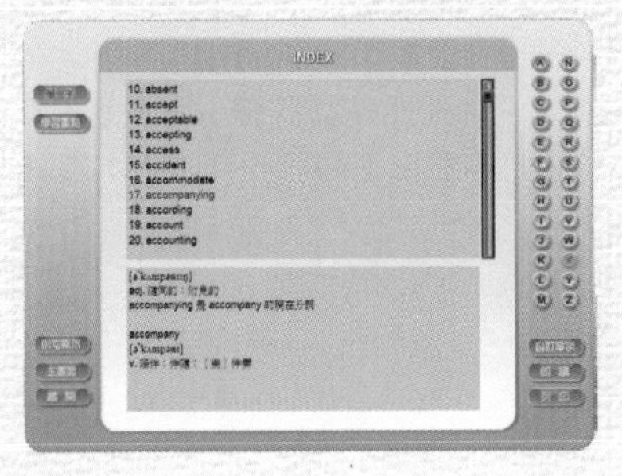

工具列說明：

❶ 單字索引：點選後可檢索全書對話的所有單字。

❷ 學習重點索引：點選後可檢索全書對話所有的學習重點。

❸ 例句顯示：選擇一單字或學習重點後再點選此鈕，會顯現該單字或學習重點的課文例句。

❹ 回主畫面

❺ 離開

❻ 字母檢索：點選後將出現以該字母開頭之單字或學習重點。

❼ 自訂單字：點選後將出現您先前所加入及編輯之自訂單字；若先前未曾自訂任何單字，則此鈕不會出現。

❽ 全文朗讀：點選後，光碟將自動依序播放視窗中所有單字或學習重點之發音，在畫面上任一處按滑鼠左鍵一下則停止。

❾ 列印：單字及學習重點解說：個別點選該單字或學習重點，會出現發音示範及其詳盡解釋；連續點選該單字或學習重點兩次，會顯現該單字或學習重點的課文例句。

課文朗讀 MP3

電腦互動光碟中含有課文朗讀 MP3 的內容，您可以放在 MP3 播放器聆聽，也可以將光碟放置於電腦中，從「我的電腦」點選您的光碟機，再從中選擇光碟資料裡 MP3 的資料夾，使用播放軟體將檔案開啟聆聽 MP3 內容。

■ 在 Vista 系統中，安裝互動光碟如遇到以下問題：

- 出現【安裝字型錯誤】之訊息。
- 出現【無法安裝語音辨識】之訊息。

請執行以下步驟：

1. 移除該產品。
2. 進入控制台。
3. 點選「使用者帳戶」選項。
4. 點選「開啟或關閉使用者帳戶控制」。
5. 將「使用（使用者帳戶控制）UAC 來協助保護您的電腦」該項目取消。
6. 再次執行安裝光碟。

■ 在 Windows 7 系統中，安裝互動光碟如遇到以下問題：

- 出現【安裝字型錯誤】之訊息。
- 出現【無法安裝語音辨識】之訊息。

請執行以下步驟：

1. 進入控制台，開啟程式集，進入程式和功能，移除該產品。
2. 進入控制台，點選「使用者帳戶和家庭安全」選項。
3. 再點選「使用者帳戶」。
4. 點選「變更使用者帳戶控制設定」。
5. 將控制拉桿調整至最底端（不要通知的位置）。
6. 按確定後，需重新啟動電腦。
7. 再次執行安裝光碟。

停看聽篇

在全球化市場環境下，e-mail 和電話是最常用的兩種溝通方式，也是許多人必須具備的職場技能。在我們學習如何面對各種不同情境的 e-mail 與電話溝通前，先來瞭解 e-mail 和電話英語的基本禮儀與必備用語，這些不但可應付各種情況，還可讓對方留下好印象。

英文 e-mail 不可不知的祕訣

I. e-mail 寫作不可不知的禮儀

❶ 主旨欄標示要清楚明確

主旨欄不可空白，並且要清楚標示來信目的，例如向對方詢問產品資訊，則最好要註明產品型號，可以採用首字大寫或標題的書寫方式，並且注意拼字是否有誤，若用全大寫或全小寫會容易讓人誤以為是垃圾信件。

收件人有時會根據主旨來判斷其重要性或決定是否閱信，因此要簡潔具體。還有，使用 urgent 或 important 這類字眼要很小心，只有在非常緊急或傳送重要訊息時才使用。

Subject: ✗

Subject: Product Information ✗

Subject: Product Information Request: Alpha Cement Mixer XC-191 ✓

Subject: Product information request: Alpha Cement Mixer XC-191 ✓

❷ 簡潔扼要

商業電子郵件宜簡單扼要，只要將書寫的目的表達清楚即可，因此一些與公事無關的內容或寒暄皆可避免。不過，寄件者與收件人之間交情不錯的話則不在此限。

商業電子郵件中應避免寒暄用語或不相關的事情：

How are you? 

How have you been? 

Hope you're having good weather. 

❸ 不要全部使用大寫字體

如果一封電子郵件全用大寫字體的話，感覺就像是 SHOUTING AT THE PERSON YOU ARE WRITING TO（對著收件人吼叫的感覺），會讓對方覺得突兀而對你的印象大打折扣，再者全大寫字體不易閱讀，會讓人無法將焦點放在信件內容上。

來比較下列這兩句話，就可以瞭解這個感覺：

I AM WRITING TO INQUIRE WHETHER YOU CAN PROVIDE YOUR LATEST CATALOG. 

I am writing to inquire whether you can provide your latest catalog.

❹ 使用主動語態而非被動語態

可以的話，盡量使用主動語態。使用被動語態的句子看起來比較像是自動回信系統發出的訊息，而用主動語態的句子比較有人情味。

來比較下列兩個句子，主動語態是不是有較積極的感覺呢？

Your order will be processed today. 

We will process your order today. 

❺ 使用中性稱呼

在不確定對方性別的情況下，應避免使用單一性別來稱呼，也就是在表達時要留意語言的政治正確性。

不確定對方性別時可以用 they 或 he/she 來取代，或稍微調整句子，避免讓女性收件人有不舒服的感覺。

The client must sign in to his account to access his order summary. 

Clients must sign in to their account to access their order summary. 

The client must sign in to access the order summary. 

Ⅱ. e-mail 不可不知的書寫架構

商業電子郵件不宜過於冗長，最好能在二、三個段落內將目的及需求表達清楚，以節省雙方的時間。

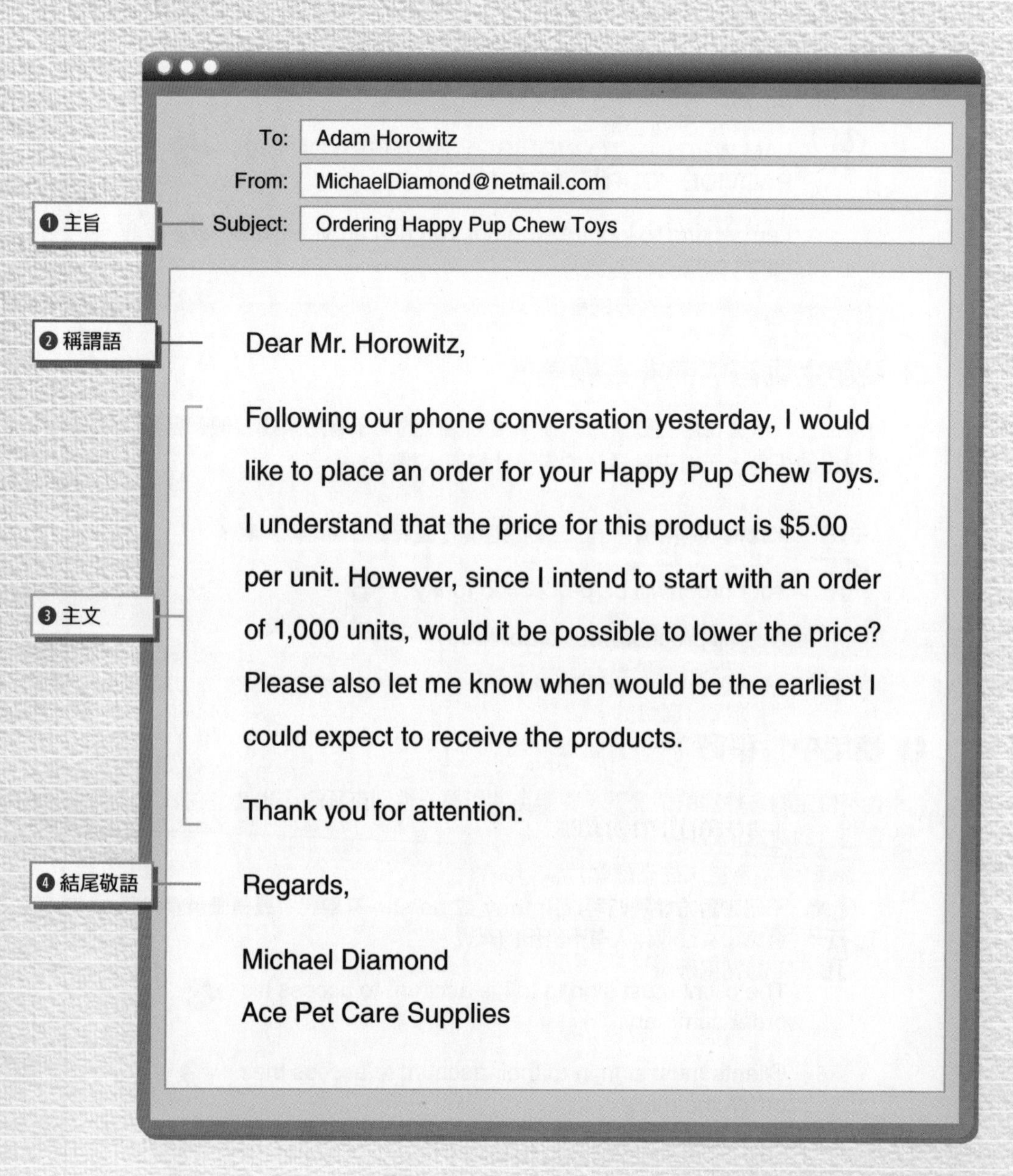

❶ 主旨（subject）

主旨欄不宜空白，需簡單說明此封信件的目的，例如：

- Holiday Notice 假期通知
- Inquiring about Tickets 詢問售票
- Prices that you requested 關於詢價
- Quote Update 最新報價
- Project Update 最新企畫

❷ 稱謂語（salutation）

商務往來或正式的書信最好要寫出稱謂，如用「Mr./Ms. + 姓氏」來稱呼對方，若與對方熟識時，可直接稱呼對方的名字即可。若不確定收信者姓名時，可用 Dear Sir or Madam 來表示。

知道收信對象是誰時	不清楚收信對象是誰時
• Dear Joe, • Hi, Helen! • Dear Mr. Brown,	• Dear Sir or Madam, • To Whom It May Concern,

❸ 主文（message body）

以簡潔為原則，通常包括開場（opening）、正文（body）、結尾（conclusion）三個部份。開場主要是表明來信目的、正文則清楚表明需求、結尾時再次表示感謝或表示願意提供協助等。（在下一頁中將介紹常見的必備句型）

❹ 結尾敬語（complimentary closing）

基本上，收信人為首次往來的客戶或廠商時，應使用正式的結尾敬語。若是經常往來的夥伴則可使用非正式的結尾敬語。

正式的結尾敬語	非正式的結尾敬語
• Sincerely yours, • Sincerely, • Best regards, • Regards, • Best wishes,	• Yours truly, • Thanks, • See you soon, • Cheers, • Take care,

Ⅲ. e-mail 書信必備句型

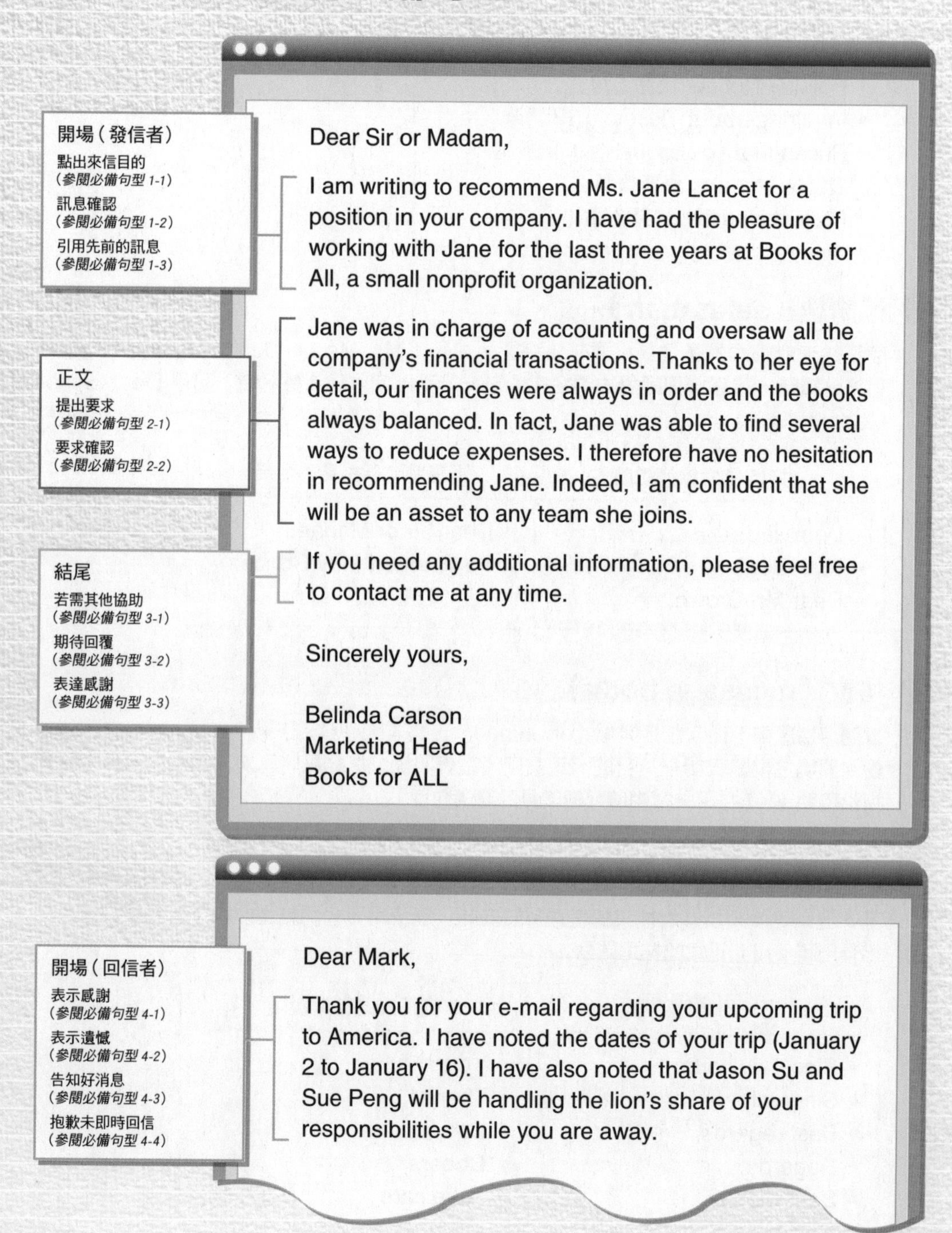

開場（發信者）
點出來信目的（參閱必備句型 1-1）
訊息確認（參閱必備句型 1-2）
引用先前的訊息（參閱必備句型 1-3）

Dear Sir or Madam,

I am writing to recommend Ms. Jane Lancet for a position in your company. I have had the pleasure of working with Jane for the last three years at Books for All, a small nonprofit organization.

正文
提出要求（參閱必備句型 2-1）
要求確認（參閱必備句型 2-2）

Jane was in charge of accounting and oversaw all the company's financial transactions. Thanks to her eye for detail, our finances were always in order and the books always balanced. In fact, Jane was able to find several ways to reduce expenses. I therefore have no hesitation in recommending Jane. Indeed, I am confident that she will be an asset to any team she joins.

結尾
若需其他協助（參閱必備句型 3-1）
期待回覆（參閱必備句型 3-2）
表達感謝（參閱必備句型 3-3）

If you need any additional information, please feel free to contact me at any time.

Sincerely yours,

Belinda Carson
Marketing Head
Books for ALL

開場（回信者）
表示感謝（參閱必備句型 4-1）
表示遺憾（參閱必備句型 4-2）
告知好消息（參閱必備句型 4-3）
抱歉未即時回信（參閱必備句型 4-4）

Dear Mark,

Thank you for your e-mail regarding your upcoming trip to America. I have noted the dates of your trip (January 2 to January 16). I have also noted that Jason Su and Sue Peng will be handling the lion's share of your responsibilities while you are away.

開場（發信者）必備句型

1-1 點出來信目的 MP3 TRACK 1

例 **I am writing to confirm** the details of my flight.
我寫信來是想確認航班的相關細節。

例 **I am writing about** the meeting we had yesterday.
我寫信來是想談談我們昨天的會議。

例 **I am writing to inquire about** accommodation availability.
我寫信來是想詢問關於住宿提供的問題。

例 **I'm interested in** placing an order for your DVD series.
我有興趣向你們訂購 DVD 系列。

例 **We'd like to invite you to** our barbecue picnic.
我們想邀請你參加我們的烤肉野餐。

例 **I am writing on behalf of** my father, who is currently in the hospital.
我寫這封信謹代表父親，他目前正在住院。

例 **I'm writing in response to** your job advertisement.
我寫這封信是回應你們的求職廣告。

1-2 訊息確認 MP3 TRACK 2

例 **I am pleased to confirm that** your application has been approved.
我很高興向你確認你的申請通過了。

例 **I confirm that** I will be attending the ceremony.
我確認我會參加典禮。

例 **Just a quick/short note to confirm that** we have received your check in the mail.
只是簡單向你確認我們已經收到信裡的支票。

例 **This is to confirm that** we have shipped your order.
這封信是確認你的訂單我們已經出貨了。

例 **We wish to confirm that** your password has been changed.
我們要確認你的密碼已經更改了。

1-3 引用先前的訊息（信件或電話內容等） MP3 TRACK 3

例 **With reference to our telephone conversation today,** I'm e-mailing you the contract to sign.
關於今天在電話裡的談話，我會用電子郵件把合約寄給你簽訂。

例 **In my previous e-mail,** I had mentioned the need to revise our plans.
在前一封電子郵件，我提到計畫有修改的必要。

例 **As I mentioned earlier about next week's plans,** we haven't yet decided what we'll do.
就像我先前提到的，關於下星期的計畫，我們還沒決定要怎麼做。

例 **As indicated in my previous e-mail,** I won't be in the office this Tuesday.
如同我前一封電子郵件提到的，我這星期二會不在辦公室。

例 **As we discussed on the phone,** I would like to close my account.
如同我們在電話裡討論的，我想要結清帳戶。

例 **From our decision at the previous meeting,** it appears that the company is making money.
從我們上一次會議的決定來看，公司似乎有賺錢。

正文必備句型

2-1 提出要求

MP3 TRACK 4

例 **We would be grateful if you could** let us know if you're coming.
如果你能讓我們知道是否會來的話，我們不勝感激。

例 **I'd appreciate it if you would** respond before Wednesday.
如果你能在星期三前回覆的話我會很感激的。

例 **I'd like to know if it would be possible** to book tickets in advance.
我想知道可不可能可以事先訂票。

2-2 要求確認

MP3 TRACK 5

例 **Please confirm that** you have received my e-mail.
請確認你已經收到我的電子郵件。

例 **Could you please confirm** that my order has been processed?
可以請你幫我確認一下我的訂單是不是已經處理了？

結尾必備句型

3-1 若需其他協助

MP3 TRACK 6

例 **For further details,** please log on to our Web site.
要知道詳細資訊的話請上我們的網站。

例 **If you wish, we would be happy to** send you our brochure.
如果你想要的話，我們很高興將文宣小冊子寄給你。

例 **Please let me know if there is anything I can do to help.**
請讓我知道是否有任何我可以幫得上忙的地方。

例 **If you have any questions, please let me know.**
如果你有任何問題，請讓我知道。

例 **If you have any further questions about** our graduate courses, **please do not hesitate to contact me.**
如果你對研究所課程還有任何問題的話，請不要客氣儘快與我聯繫。

3-2 期待回覆

MP3 TRACK 7

例 **I look forward to your reply.**
我期待您的回覆。

例 **I hope to hear from you soon.**
我希望不久就能得到你的消息。

例 **Please respond at your earliest convenience.**
請儘速回覆。

例 **Feel free to contact me by** phone or e-mail.
隨時透過電話或電子郵件的方式與我聯繫。

3-3 表達感謝

MP3 TRACK 8

例 **Thank you for your help.**
感謝您的協助。

例 **Any comments will be much appreciated.**
任何意見都將非常感謝。

例 **Thank you again for choosing** our company.
再次感謝您選擇我們公司。

例 **Thank you for taking this into consideration.**
感謝您將此列入考慮。

例 **I appreciate any feedback you may have.**
您的任何意見回饋我都十分感謝。

例 **Your help with this would be much appreciated.**
您對此事的協助我們將萬分感謝。

例 **Your prompt attention to this matter will be appreciated.**
您對此事立即的處理我們將感激不盡。

開場（回信者）必備句型

4-1 表示感謝

MP3 TRACK 9

例 **Thank you for** purchasing our product.
謝謝您購買我們的產品。

例 **Thank you for your query about** our English summer camp program.
謝謝您詢問我們英文夏令營的計畫。

例 **Thank you for your reply.**
感謝您的回覆。

例 **Thank you for your interest in** our store.
謝謝您對本店感興趣。

例 **Thanks for getting in touch with us.**
謝謝您與我們聯繫。

4-2 表示遺憾

MP3 TRACK 10

例 **We regret to inform you that** your loan application has been denied.
我們很遺憾通知您的貸款申請被拒絕了。

例 **It is with regret / great sadness that** we share the news of our grandmother's death.
知道奶奶過世的消息，我們都同感悲傷。

例 **After careful consideration, we have decided** to sell our house.
經過再三考慮後，我們決定賣掉我們的房子。

例 **I'm sorry to hear that** you won't be coming this weekend.
聽到你這週末無法前來，我很遺憾。

例 **I'm sorry to say that** you did not win the prize.
很遺憾要跟你說，你並未贏得獎項。

例 **Unfortunately,** I don't have this information. I'll need to get back to you.
很遺憾，我沒有這項資訊，我得之後再回覆你。

4-3 告知好消息

MP3 TRACK 11

例 **It is my pleasure to tell you** about our new products.
能告訴你我們的最新產品是我的榮幸。

例 **We are delighted to let you know** that your shipment will go out tomorrow.
我們很高興告知你訂單明天就會出貨。

例 **We are excited to inform you** of our latest offers.
我們很興奮要通知你我們最新的出價。

4-4 抱歉未即時回信

MP3 TRACK 12

例 **Sorry for the late reply.**
抱歉回信晚了。

例 **Sorry for the delay in responding to** your inquiry.
抱歉這麼晚回覆你的問題。

電話英語不可不知的祕訣

I. 電話模擬情境圖

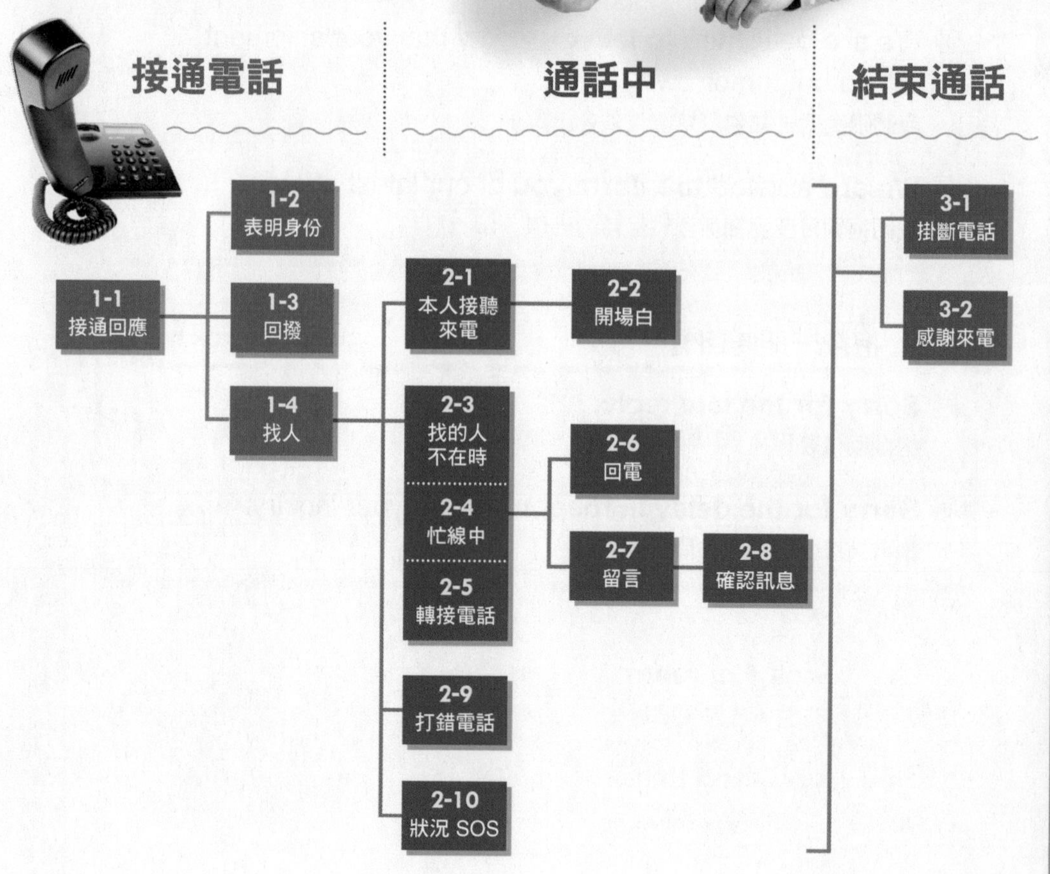

接通電話

1-1 接通回應

MP3 TRACK 13

非正式	• Hello? 喂？
正式	• Doctor Lee's office. 李醫師診所。 • Thank you for calling ABC Semiconductors. Selina speaking. How may I help you? 感謝您致電 ABC 半導體公司。我是莎琳娜。有什麼需要效勞的嗎？

1-2 表明身份

MP3 TRACK 14

非正式	• Hey Ellen. **It's** Mary **calling.** 你好，艾倫。我是瑪莉。
正式	• Hello, **this is** Shelly Chen **calling from** Delta Designs. 你好，我是三角設計公司的陳雪莉。 • Hello, **this is** Mary **calling on behalf of** Mr. Lin. 哈囉，我是代表林先生打來的瑪莉。 • **I'm calling from** Taipei/Tokyo/Hong Kong/Beijing. 我是從台北／東京／香港／北京打來的。

1-3 回撥

MP3 TRACK 15

- Hello, Mr. Scott. **I'm returning your call.**
  您好，史考特先生。我是要回電給您。
- Ms. Smith, **you tried to reach me earlier**?
  史密斯小姐，妳之前要找我嗎？

- **I'm sorry I missed your call** this morning.
  抱歉，今天早上沒接到你的來電。
- **I believe someone from this number called me.**
  我想有人打這支電話找我。

1-4 找人

MP3 TRACK 16

非正式	• **Is** Kate **in**? 凱特在嗎？ • **Is** Billy **there, please**? 請問比利在嗎？ • **Can I talk to** your sister? 可以請姊姊來接電話嗎？
正式	• **May I speak with** Mr. Oldman, **please?** 請問歐德曼先生方便接電話嗎？ • **Would** the doctor **be in/available?** 請問醫生現在有空嗎？ • **Could you please put me through to** Tom? 你可以幫我轉接給湯姆嗎？ • **May I have extension** 432? 可以幫我轉接分機四三二嗎？

接通電話

2-1 本人接聽來電

MP3 TRACK 17

- **This is** she/he. 我就是。
- **Speaking.** 我就是。
- **This is** Mark **speaking.** 我就是馬克。

2-2 開場白

MP3 TRACK 18

- **Is this a convenient time to talk?**
  現在方便說話嗎？
- **Would you have a moment? I wanted to talk to you about** plans for this afternoon.
  你現在有空嗎？我想要和你談談今天下午的計畫。
- **I'm calling to** make a reservation.
  我打電話來是要訂位。
- **I'd like to talk to** Martha **about** the business trip next month.
  我想要找瑪莎談下個月出差的事。

電話禮儀

打電話找人時，除了說明目的外，最好先禮貌詢問對方此時是否方便說話，對方可能正在開會或有急事要去處理，事先詢問可以展現你的禮貌與體貼。

2-3 找的人不在時

MP3 TRACK 19

- Tim's **not in. Can you call back later**?
  提姆不在。你可以晚一點再打來嗎？
- **I'm afraid he's away from his desk.**
  恐怕他不在位置上。
- Jason **is not available at the moment**. May I ask who's calling?
  傑森現在無法接聽電話。請問是哪裡找嗎？

上班族加油站

要找的人不在時可以說 be not in 或 be away from one's desk，也可以用 unavailable 來表明無法接聽電話。

2-4 忙線中

MP3 TRACK 20

- Tina **is on another call right now.**
  蒂娜現在正在講電話。
- Mr. Watson **is on another line at the moment**. May I place you briefly on hold?
  華特森先生目前正在忙線中。可以請您稍待一下嗎？

上班族加油站

表達「電話忙線中」，你可以用 on the phone、on another call/line 或 the line is busy/engaged 來表示。

電話禮儀

如果對方要找的人正在忙線中，別忘了先詢問對方是否願意稍等，或請對方稍後再來電。

2-5 轉接電話

MP3 TRACK 21

非正式	• **Hang on one second, please.** 請稍等一下。 • **The line's free now. I'll put you through.** 電話目前沒佔線。我幫你轉過去。 • **I'll connect you now. / I'm connecting you now.** 我將為你轉接。 • **Just a sec. I'll get** him. 稍等一下。我轉給他。
正式	• **Would you mind holding while I put you through?** 我替您把電話轉過去，請您稍等一下好嗎？ • **One moment please.** 請稍等一下。

上班族加油站

「轉接」除了可以用 put sb through、connect sb 之外，也可以用 transfer、switch sb to 來表示。

上班族加油站

「回電給某人」可以用 call/phone sb back 或 return sb's call 來表示。

2-6 回電

MP3 TRACK 22

請對方要找的人稍後回電	• Henry is on another line right now. **May I ask him to call you back?** 亨利正在忙線中。要我請他回電給你嗎？ • **May I ask** Peter **to return your call?** 要我請彼特回電給你嗎？
表示晚一點再回電	• **No, that's fine. I'll call back.** 不用了，沒關係。我會再打來。 • **When would be the best time to call again?** 什麼時候再打來最合適呢？

2-7 留言

MP3 TRACK 23

表示要幫忙留言	• Mr. Jackson is in a meeting at the moment. **May I take a message**? 傑克森先生目前正在開會。要我留言嗎？ • I'm afraid he's stepped out. **Would you like to leave a message**? 恐怕他出去了。您要留言嗎？
請留言給要找的人	• Yes, **can you tell** him his brother **called, please.** 好的，請告訴他是他弟弟打來的。 • Thanks. **Could you ask him to call** Bob Morgan **when he gets in**? 謝謝，他回來的時候可以請他打給鮑伯．摩根嗎？ • **Do you have a pen handy**? **I don't think he has my number.** 你手邊有筆嗎？我想他應該沒有我的電話。

- **Could you please have him call me back** on my cell phone? The number is 0975-476-788.
  可以麻煩他回電打手機給我嗎？號碼是 0975-476-788。

電話禮儀

在職場上要幫忙留言時，可簡單說明一下要找的人無法接聽的原因。

2-8 確認訊息

MP3 TRACK 24

- **OK, I've got it all down.** 好的，我都記下來了。
- **Let me repeat that just to make sure.** 讓我重複一遍再做確認。
- **Did you say** 353 Bade Rd.? 你是說八德路三百五十三號嗎？
- **I'll make sure he gets the message.** 我會確認他收到訊息。

2-9 打錯電話

MP3 TRACK 25

- **I'm afraid you have the wrong number.** 恐怕你打錯電話了。
- **There's nobody here by that name.** 這裡沒有叫那個名字的人。
- **There is no such person at this number.** 這支電話沒有這個人。

2-10 狀況 SOS

MP3 TRACK 26

- **Could you speak a little slower, please? My English isn't very strong.**
  可以請你說慢一點嗎？我的英文不是很強。
- Could you please call me back? **I think we have a bad connection.**
  可以請你回撥給我嗎？通訊品質不太好。

- **Could you please repeat that?** 可以請再重複一遍嗎？
- **Would you mind spelling that for me?**
  你介意幫我拼一下那個字嗎？

上班族加油站

聽不清楚的狀況包括：bad connection/reception（收訊不良）、有 echo（回音）、interference（干擾）等情況。

- **Could you speak up a little, please?**
  可以請你說大聲一點嗎？

電話禮儀

使用 could 來代替 can 語氣上會較為客氣，再加上 please 會讓句子聽起來更有禮貌。

結束通話

3-1 掛斷電話

MP3 TRACK 27

- **I have another call coming through. I better run.**
  我有另一通電話進來。我得掛了。
- **I'm afraid that's my other line.** 恐怕我得接另一通電話。
- **I have to let you go now.** 我得掛你電話了。
- **I'll talk to you again soon. Bye.** 我們再聯絡。再見。
- **May I call you back in ten minutes?** 我可以十分鐘後打給你嗎？

3-2 感謝來電

MP3 TRACK 28

- **Thank you for calling.** 感謝你的來電。
- **Nice talking to you.** 很高興與你通話。

Ⅱ. 電話答錄篇

電話答錄機留言

MP3 TRACK 29

Hello. You've reached 2957-3489. Please leave a detailed message after the beep. Thank you.
您好，您撥的電話是 2957-3489。請在嗶一聲之後留下詳細的訊息。謝謝。

Hi, this is Sandra. I'm sorry I'm not available to take your call at this time. Leave me a message and I'll get back to you as soon as I can.
您好，我是珊卓拉。很抱歉現在無法接聽您的電話。請留言，我會儘快與您聯絡。

來電者留言

非正式

Hey Richard. It's Jack. Call me!
嗨理查。我是傑克，打電話給我！

正式

Hello, this is Mike calling for Luke. Could you please return my call as soon as possible? My number is 2561-5689. Thank you.
你好，我是麥克要找路克。你可以儘快回電給我嗎？我的電話是 2561-5689。謝謝。

公司語言信箱

Thank you for calling Livetech. If you know the extension number of the person you wish to talk to, please enter the number now, followed by the pound key. Otherwise, please choose from the following menu: to speak to a sales representative, press 1; for technical assistance, press 2; for billing, press 3; for general inquiries, press 0. To listen again to the menu, press star.

謝謝您來電 Livetech 科技公司。如果您知道所找的人的分機號碼，請直撥分機號碼再撥井字鍵。否則請選擇下列的功能選單：業務代表請按一；技術協助請按二；繳費請按三；一般諮詢請按〇；重聽選單請按星號。

Ⅲ. 電話號碼怎麼說才正確？

MP3 TRACK 30

電話號碼上的數字鍵通常個別唸出。

212-559-6970

area code two one two,
five five nine,
six nine seven oh.

2 後四碼唸的時候也可以「兩位數」為單位，不過一個一個唸還是較常見。

516-848-7789

- five one six, eight four eight, seventy-seven, eighty-nine
- five one six, eight four eight, seven seven eight nine（較常見）

3 美國免付費電話的區域號碼是 800 或 900，唸法如下：

1-800-564-5544

one eight hundred, five six four, five five four four.

在電視廣告或廣播中的電話號碼為連號時，通常不會個別唸出，不過一般人講電話號碼時，直覺上還是會一個一個唸。

416-888-1600

- four one six, triple eight, one six hundred（較常出現在商業廣告中）
- four one six, eight eight eight, one six oh oh（大多數人還是會這樣講）

人際關係篇

職場上除了一般業務往來之外，建立人際友好關係也很重要，不僅可以拉近彼此之間的距離，增加信任感，合作起來也會更順暢愉快。好的人際關係首重真誠，無論是祝賀、感謝、表達關心或弔唁，都要盡快將你的心意傳達給對方，才能讓對方感受到你的誠意。

UNIT

01

請假通知

Communicating a Leave of Absence

I. 請假通知一定要會的單字片語

MP3 TRACK 31

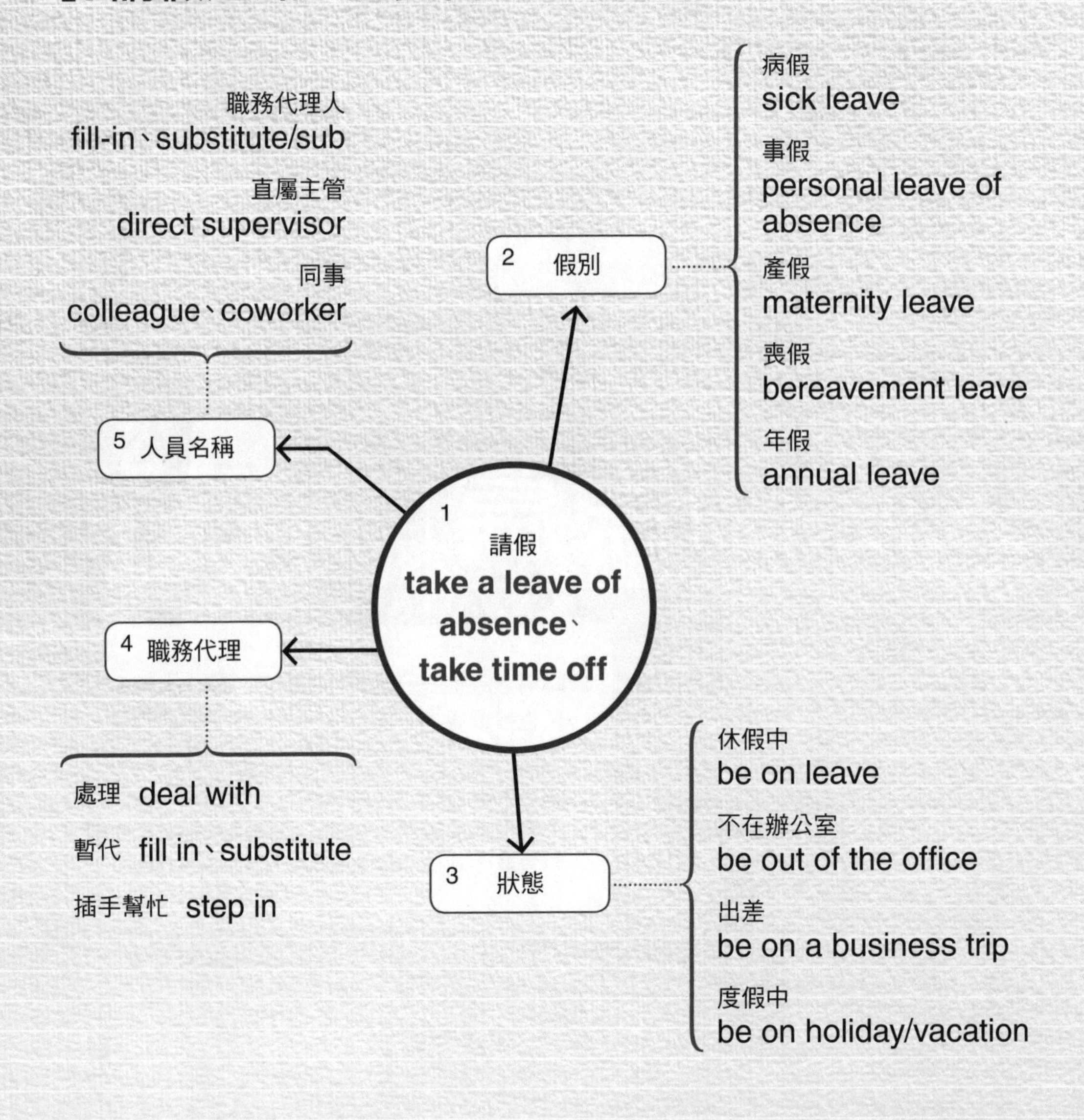

Ⅱ. 請假通知一定要會的句型

句型 1 告知請假日期

I will be out of the office for + 一段時間 + **starting** + 日期

例 I will be out of the office for two weeks starting January 2.
從一月二日開始我將有兩個星期不在辦公室。

句型 2 交代目前工作

I am currently working on sth（工作項目）

例 I am currently working on a sales presentation for the Peterson Group.
我目前正在進行彼得森集團的業務簡報。

句型 3 告知職務代理人

S. **will be stepping in to handle my responsibilities.**

例 Jason Su will be stepping in to handle my responsibilities.
蘇傑森將會插手處理我負責的工作。

句型 4 緊急聯絡方式

You may contact me + 聯絡方式 + 時間

例 You may contact me by cell phone at anytime.
你可以隨時打手機與我聯絡。
→ 也可以用 via e-mail 哦！

Ⅲ. 如何用 e-mail 請假

To: Jane Anderson

Subject: Upcoming Leave of Absence

Dear Jane,

請假時間

I am writing this quick e-mail to confirm with you that I will be out of the office for two weeks starting January 2. As I had mentioned to you earlier, I will be traveling to America for a family reunion[1] during this time. I will be back in the office on Monday, January 16.

交代目前工作

交代職務代理人

I am currently working on a sales presentation for the Peterson Group, and Jason Su will be stepping in to handle my responsibilities. I am also involved in the preparations for the upcoming Design Expo in February, and Sue Peng will be responsible for these matters while I am away.
In addition, I've talked to the other members of my team, and they are also ready to step up to the plate during my absence.[2]

緊急聯絡方式

While I am in America, you may contact me by cell phone at anytime. I will also be available via e-mail. My main contact in America will be my brother, Tim Lee, and his home phone number is (219) 462-6499.

Thank you once again for all of your assistance. I look forward to returning to the office on Monday, January 16.

Sincerely,

Mark Lee

上班族加油站

step up to the plate 表示「負責做某事」，通常帶有事情可能不容易處理的意思。

Vocabulary & Phrases

1. reunion [riˋjunjən] *n.* 團聚
2. absence [ˋæbsəns] *n.*
   不在；缺席

中文翻譯

收件人：珍．安德森
主旨：即將要請假

親愛的珍：

我匆匆寫下這封電子郵件是要跟妳確認，從一月二日開始我將有兩個星期不在辦公室。如同我先前跟妳提過的，這段時間因為家人團聚的關係我會去美國。我會在一月十六日星期一回到辦公室。

我目前正在進行彼得森集團的業務簡報，蘇傑森將會插手幫忙處理我負責的工作。我也參與了即將在二月舉辦的設計展覽會的準備工作，我不在時彭蘇會負責這些事情。除此之外，我也和其他組員談過，在我請假期間，他們也準備好會負責處理這些工作。

我在美國的時候，妳可以隨時打手機給我，也可以用電子郵件跟我聯絡。我在美國的主要聯絡人是我弟弟李提姆，他家的電話號碼是 (219) 462-6499。

再次感謝妳的協助。我很期待一月十六日星期一回到辦公室上班。

謹上

李馬克

文化補給站

如果你的業務聯絡窗口將私人聯絡電話留給你，並且告訴你可以在任何時候打電話給他的話，就表示他不介意自己的私人行程被打擾，但若非緊急事件還是盡量不要打擾比較好。

如果對方未留下聯絡方式，那麼最好是根據他留下的電子郵件訊息，按照裡面的指示來處理。

以下是休假通知的範例：

I will be out of the office **on holiday / on a business trip / for personal reasons** from November 7 to November 14. I will review your message as soon as I get back on November 15. If you need assistance, please contact [name] at [e-mail address] or call her at [phone number]. Thank you for your understanding.

Ⅳ. 如何用 e-mail 回覆請假信

To: Mark Lee

Subject: Re: Upcoming Leave of Absence

Dear Mark,

Thank you for your e-mail regarding[1] your upcoming trip to America. I have noted the dates of your trip (January 2 to January 16). I have also noted that Jason Su and Sue Peng will be handling the lion's share of your responsibilities while you are away.

If any issues should arise,[2] I will be sure to contact you on your cell phone or via e-mail. I have also received your brother's emergency contact information in America.

Once again, thank you for confirming the dates of your upcoming trip with me. Have a wonderful time in America.

Yours truly,

Jane Anderson

上班族加油站

lion's share 這個片語是從寓言故事裡獅子分配到大部分的獵物引申而來，表示「大部分的、幾乎全部的」。

Vocabulary & Phrases

1. regarding [rɪˋgɑrdɪŋ] *prep.* 關於
2. arise [əˋraɪz] *v.* 出現；發生

中文翻譯

收件人：李馬克
主旨：回覆：即將要請假

親愛的馬克：

非常謝謝你來信提到美國行一事。我已經記下你的旅遊日期（一月二日至一月十六日）。我也記下蘇傑森和彭蘇在你不在期間，會幫忙處理你大部分的工作。

如果發生任何事情的話，我一定會打手機或是用電子郵件來跟你聯絡。我也有收到你弟弟在美國的緊急聯絡資訊。

再次感謝你和我確認即將去旅行的日期。祝你在美國玩得愉快。

謹上

珍．安德森

Try it! 換你試試看!

1. 我想確認從五月一日開始我會不在辦公室。

2. 我目前正在進行銷售部門的每季預算審查。

(每季預算 quarterly budget)

3. 我不在這段期間，我同事 Joe Hu 會插手幫忙處理我負責的工作。

(插手 step in)

4. 萬一有緊急狀況，你可以打下面的電話與我聯絡：(312) 339-4558。

答案請參閱第 366 頁

V. 用電話請假一定要會這樣說

MP3 TRACK 32

告知請假時間

A: Will you be available to go over the new sales plan next Tuesday?

你下星期二有空看一下新的銷售計畫嗎？

B: Sorry, but **I will be out of the office for** one week **starting** next Monday.

抱歉，從下星期一開始我會有一個星期不在辦公室。

A: **I'm calling to remind you that I will be away** on a business trip next week.

我打電話是要提醒你，下個星期我會去出差。

B: Thank you for letting me know about this.

謝謝你讓我知道這項訊息。

詢問職務代理人

A: **Who will be handling** the Thompson account **while you are gone?**

你不在的時候湯普森客戶會由誰來處理呢？

B: Ed Wu will be responsible for this account while I am out of the office.

我不在的時候吳艾德會負責這位客戶。

告知職務代理人

A　Cindy Ho **will be filling in for me while I'm on vacation.**

我去度假的時候，何辛蒂會代理我的職務。

B　Sounds good. Be sure to bring her up to speed on all your projects before you go.

聽起來不錯。在你離開之前務必要讓她瞭解你所有案子最新的狀況。

詢問緊急聯絡方式

A　**In a worst-case scenario, how can we get in touch with you?**

在最糟的情況下，我們要怎麼與你聯繫？

B　You can always get in touch with me at my parents' house. The number there is (444) 602-8904.

你隨時可以在我父母家聯絡上我。那裡的電話是 (444) 602-8904。

告知緊急聯絡方式

A　**In case of an emergency, please contact me at** the Bayside Hotel—(02) 344-7684.

萬一有緊急狀況的話，請打到海灣飯店與我聯繫：(02) 344-7684。

B　I'm sure everything will be fine, but it's nice to have this information, just in case.

我相信一切都會沒問題，但有這個資訊不錯，只是以防萬一。

VI. 用電話説明請假交接事宜

優先科技的業務代表吳珍在請假前打電話告知重要客戶，交代工作交接事宜。

LISTENING 請聽 MP3 TRACK 33 SPEAKING 請跟著 MP3 唸唸看

Jane: Hello? Mr. Brown? This is Jane Wu from AdvanceTech.

Mr. Brown: Hi, Jane. What can I do for you?

請假時間

Jane: Well, I just wanted to let you know that I will be taking a leave of absence from the office for two weeks.

Mr. Brown: Thanks for informing me. When will you be gone?

Jane: I'll be leaving on January 30, and I will return to the office on February 13.

Mr. Brown: All right. Who will be filling in for you while you're away?

告知職務代理人

Jane: My colleague, Fiona Wang, will be handling my duties for me while I am away.

Mr. Brown: And in case of an emergency, how can we get in touch with you?

緊急聯絡方式

Jane: You can always reach me on my cell phone or by e-mail. I've also e-mailed you an emergency contact number, along with the exact dates of my trip.

Mr. Brown: It sounds as if you've thought of everything then. Thanks again, Jane, for keeping me informed of your plans.

你好？布朗先生嗎？我是優先科技的吳珍。

嗨，珍。有什麼我可以幫妳的嗎？

嗯，我只是要讓你知道我將會請假兩個星期。

謝謝妳通知我。妳什麼時候要離開？

我會在一月三十日離開，然後在二月十三日回到辦公室。

好的。妳不在的時候誰會代理妳的職務？

我不在的時候，我同事王菲歐娜會替我處理。

那萬一有緊急狀況的話，我們要怎麼跟妳聯絡？

你隨時可以透過手機或是電子郵件找到我。我也已經將緊急聯絡電話，連同我旅行的詳細日期寄電子郵件給你了。

聽起來妳似乎已經把所有事情都設想好了。珍，再次感謝妳告知我妳的計畫。

Try it! 換你試試看！

WRITING 請依提示寫出完整句子	☐
LISTENING 請聽 MP3 TRACK 34	☐
SPEAKING 請跟著 MP3 唸唸看	☐

1. Ⓐ When will you be out of the office?

   Ⓑ ______________________________

   ____

   (one week / next Monday)

2. Ⓐ What are you working on at this moment?

   Ⓑ ______________________________

   ____

   (work on / analysis of overseas branches)

   Ⓐ Please remember we need to have this completed before the end of the month.

3. Ⓐ Who will be handling your responsibilities while you are out of the office?

   Ⓑ ______________________________

   ____

   (colleague / Doris Zhou / fill in / away)

4. Ⓐ I don't anticipate any problems, but how should we contact you if there is an emergency?

   Ⓑ ______________________________

   ____

   (emergency / my mother's house / (312) 443-6786)

答案請參閱第 366 頁

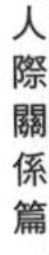

UNIT
02 表達祝賀
Expressing Congratulations

I. 表達祝賀一定要會的單字片語

MP3 TRACK 35

蒸蒸日上
continued growth/success in the future

幸福美滿
happiness in the years ahead

白頭偕老
long and healthy life together

事事順利
all the best for the future

結婚 wedding

訂婚 engagement

生子 birth

結婚紀念日 wedding anniversary

升官 promotion

畢業 graduation

生日 birthday

開幕；就職 inauguration

Ⅱ. 表達祝賀一定要會的句型

句型 1　寫信道賀

I am writing to congratulate sb + **on** sth!

例 I am writing to congratulate you on your recent wedding!
我寫信是要向你新婚賀喜。

句型 2　表達高興的心情

I was pleased + to V.

例 I was pleased to hear this great news from my colleague.
我從同事那裡聽到這個好消息時感到很高興。

句型 3　表達祝賀

I wish you + 祝賀語

例 I wish you both nothing but happiness in the years ahead.
我希望你們兩個將來能過著幸福的日子。

Ⅲ. 用 e-mail 祝賀新生兒誕生與回覆

To: Mary Oliver

Subject: Congratulations on Your New Baby!

Dear Mary,

I know that this must be a busy time for you. However, I wanted to write to congratulate you on the birth of your son this past weekend.

恭喜喜獲麟兒

I've been looking forward to hearing this good news for months. We've all been on pins and needles as the day grew closer. And now that your son is here, we are all so excited.

表達高興的心情

Motherhood is certainly very special, and I am certain that you will be a wonderful mother.

新手媽媽祝賀語

Once again, congratulations, and enjoy the new addition to your family!

Sincerely,

Cindy Huang

中文翻譯

收件人：瑪莉・奧莉維
主旨：恭喜喜獲麟兒！

親愛的瑪莉：

我知道這段時間妳一定很忙。然而，我還是要寫信祝賀妳上個週末喜獲麟兒。

好幾個月來我一直期待聽到這個好消息。日子一天天接近，我們也如坐針氈。現在妳兒子出生，我們全都感到無比興奮。

當了媽媽一定很不一樣，我很確定妳會是一個很棒的母親。

再次恭喜妳，並且好好享受家裡新成員帶來的樂趣！

謹上

黃辛蒂

To: Cindy Huang
Subject: Re: Congratulations on Your New Baby!

Dear Cindy,

Thank you so much for your heartfelt e-mail. It was so nice to read your kind words.

Yes, it has been a busy time. But it has also been wonderful. Nothing is better than holding your own child in your arms as he sleeps.

Thank you again, Cindy, for your e-mail. I can't wait for you to meet my son.

Sincerely,

Mary Oliver

中文翻譯

收件人：黃辛蒂
主旨：回覆：恭喜喜獲麟兒！

親愛的辛蒂：

非常感謝妳令人窩心的來信。收到妳貼心的話語真的很高興。

是的，這段時間真的很忙，但是也很不可思議。沒有什麼比抱著自己熟睡的孩子更棒的了。

辛蒂，再次感謝妳的來信。我等不及要讓妳見見我兒子了。

謹上

瑪莉．奧莉維

Ⅳ. 用 e-mail 祝賀得獎與回覆

To: Tony Jenkins
Subject: Congratulations on Your Award!

Dear Tony,

恭喜得獎

I am writing this brief e-mail to offer my heartiest congratulations to you on the award that was given to you last weekend. It certainly is quite an honor to be named the Salesperson of the Year for your entire company!

表達高興的心情

I was pleased to hear this great news from my colleague, Simon Zhang, who was at the awards banquet. I can't say that I was surprised to hear that you'd won, though, since you've always offered such great sales service.

得獎祝賀語

It's nice to know that I will be working with such a top salesperson, and I wish you many years of continued success in the future. Once again, congratulations, and enjoy your award!

Sincerely,

Harris Wu

中文翻譯

收件人：東尼．傑金斯
主旨：恭喜你得獎！

親愛的東尼：

我寫這封短信是要為你上星期獲獎獻上我誠摯的祝賀。獲選為全公司年度最佳業務員無疑是一項榮耀！

我很高興從我同事張賽門那裡得知這個大消息，他當時就在頒獎宴上。不過，聽到你得獎我並不感到驚訝，因為你一直都提供非常棒的銷售服務。

知道我將會繼續和這麼頂尖的銷售員合作感到很高興，也祝福你未來能蒸蒸日上。再次恭喜你，並好好享受得獎的喜悅吧！

謹上

吳哈利斯

To: Harris Wu

Subject: Re: Congratulations on Your Award!

Dear Harris,

Thank you so much for your e-mail, and I'm sorry that I have not been able to get back to you sooner.

Actually, I was very surprised to win this award, since I work with such a great team of salespeople. It is quite an honor, though, and I am very grateful to have received this title.

Thank you once again, Harris, for taking the time to congratulate me on this award. I look forward to our continued cooperation in the future.

Sincerely,

Tony Jenkins

中文翻譯

收件人：吳哈利斯

主旨：回覆：恭喜你得獎！

親愛的哈利斯：

非常感謝你的來信，很抱歉沒能早點回信給你。

事實上，得到這個獎我感到非常驚訝，因為和我共事的是一群優秀的業務團隊。然而，還是感到相當榮幸，我很感謝能獲得這項頭銜。

哈利斯，再次感謝你特地祝賀我獲獎。我也期待我們往後持續的合作。

謹上

東尼‧傑金斯

V. 用 e-mail 祝賀新婚與回覆

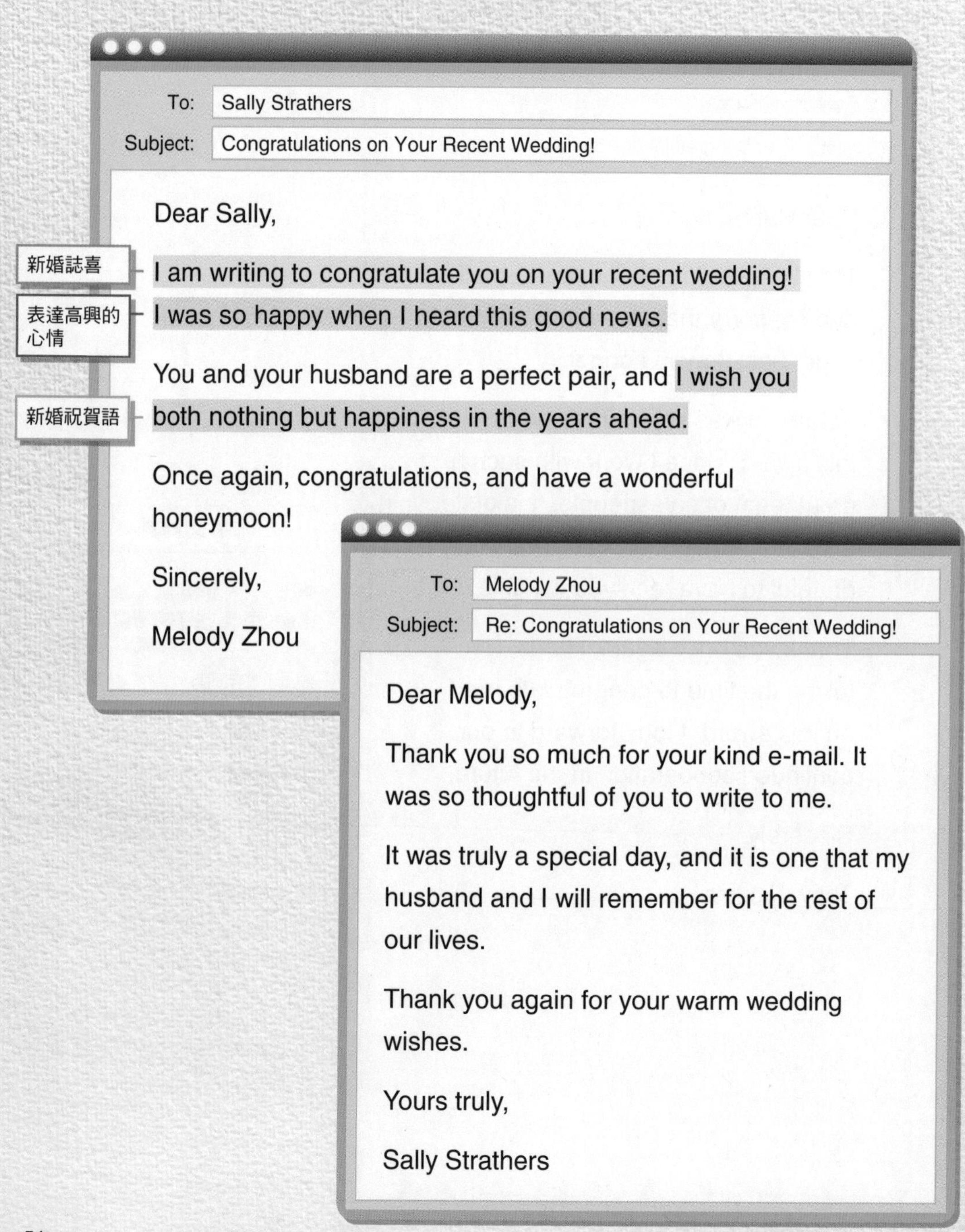
To: Sally Strathers
Subject: Congratulations on Your Recent Wedding!

Dear Sally,

新婚誌喜 — I am writing to congratulate you on your recent wedding!
表達高興的心情 — I was so happy when I heard this good news.

You and your husband are a perfect pair, and I wish you both nothing but happiness in the years ahead. — 新婚祝賀語

Once again, congratulations, and have a wonderful honeymoon!

Sincerely,

Melody Zhou

To: Melody Zhou
Subject: Re: Congratulations on Your Recent Wedding!

Dear Melody,

Thank you so much for your kind e-mail. It was so thoughtful of you to write to me.

It was truly a special day, and it is one that my husband and I will remember for the rest of our lives.

Thank you again for your warm wedding wishes.

Yours truly,

Sally Strathers

中文翻譯

收件人：莎麗．史翠瑟思
主旨：新婚賀喜！

親愛的莎麗：

我寫信是要為向妳新婚賀喜！當我聽到這個好消息時真的很開心。

妳和妳先生是天造地設的一對，我希望你們將來能過著幸福美滿的日子。

再次恭喜妳，並祝你們蜜月愉快！

謹上

周美樂蒂

收件人：周美樂蒂
主旨：回覆：新婚賀喜！

親愛的美樂蒂：

非常感謝妳如此體貼的來信。妳真的很貼心還寫信給我。

那真的是很特別的一天，也是我和我先生往後人生會永遠記得的一天。

再次感謝妳溫暖的婚禮祝福。

謹上

莎麗．史翠瑟思

Try it! 換你試試看!

1. 我要為你的新婚獻上祝賀。

____

____

____

2. 恭喜你最近得獎。

____

____

____

3. 我很高興聽到這個好消息。

____

____

____

4. 我祝福你未來蒸蒸日上。

____

____

____

答案請參閱第 366 頁

VI. 用電話祝賀一定要會這樣説

MP3 TRACK 36

祝賀晉升

A	**I'm calling to congratulate** you **on** your promotion to vice-president.	我打電話是要祝賀你升上副總經理。
B	Thanks. I wasn't expecting to get it.	謝謝。我沒有預期會獲得升遷。

祝賀交易達成

A	**I heard through the grapevine that you made a big sale** this morning.	我聽到謠傳你今天早上完成一筆大交易。
B	Thanks, I did. It's nice that everything worked out.	謝謝，沒錯，事事順利還真的不錯。

從 grapevine（葡萄藤）聽來的，表示不是正式發布的消息，而是聽來的小道消息。

A	**Did you really land** the DuoTech account **today?**	你今天真的獲得二重科技這個客戶了嗎？
B	I did. I've been working for months to get their business.	是的。我一直努力了好幾個月才接到他們的生意。

在口語用法中，得到工作機會可以用 land 這個動詞來表示，例如 land a job、land a role。

婚禮祝福

A You and Sue **make such a great couple. It was nice to hear that you finally tied the knot**.

你和蘇多麼的登對。聽到你們終於共結連理真是太好了。

B Thanks. We had a wonderful wedding.

謝謝。我們有個很棒的婚禮。

A Hey, **I've been looking for the chance to congratulate you on** your wedding. **Congratulations!**

嘿，我一直找機會要向妳新婚賀喜。恭喜！

B Thanks. It was the most magical day of my life.

謝謝。那是我生命中最奇妙的一天。

為新生兒祝賀

A **Congratulations on the birth of** your daughter. How does it feel to be a father?

恭喜你生了女兒。當爸爸的感覺如何？

B It feels great. It's amazing to look into my little girl's eyes.

感覺很棒。看著小女兒眼睛感覺很奇妙。

A Nice to see you back at work. **Congratulations on the birth of your boy.**

很高興看到你回來工作。恭喜你生了兒子。

B Thank you. He's such an incredible little guy.

謝謝。他是個令人不可思議的小男孩。

Ⅶ. 用電話表達祝賀

艾賽特合夥人公司的賴珍妮打電話為他們事業夥伴獲獎獻上祝賀。

LISTENING 請聽 MP3 TRACK 37 ☐ SPEAKING 請跟著 MP3 唸唸看 ☐

Jenny: Hi. This is Jenny Lai at Exeter Associates. Is Mark Lee in?

Mark: Speaking. What's on your mind, Jenny?

恭喜 — Jenny: I'm calling to congratulate you on the award you won last night.

Mark: Wow. Word sure does travel fast.

Jenny: Well, good news travels fast, Mark, and this is great news!

Mark: To be completely honest, I never expected to be named manager of the year. I didn't even know I was in the running.

Jenny: Come on, Mark. With the outstanding year your department's been having, this couldn't have come as a surprise.

Mark: Well, it did. I'm just lucky to have such great team members working for me.

祝賀語 — Jenny: And we at Exeter are lucky to be working with an award-winning manager like you. Congratulations once again!

Mark: Well, thank you once again, Jenny. I appreciate your taking the time to call and congratulate me.

嗨。我是艾賽特合夥人公司的賴珍妮。李馬克在嗎？

我就是。珍妮，有什麼事嗎？

我是要打電話恭喜你昨天晚上獲獎的事。

哇，消息傳得真快。

嗯，馬克，好事傳千里，而且這是個大消息！

老實說，我從沒有料到會被選為年度管理人。我甚至不曉得我在候選名單中。

拜託，馬克。你們部門這一年表現這麼傑出，這不是什麼太意外的事。

嗯，確實是。我只是很幸運有這麼棒的團隊跟我共事。

而我們艾賽特則是很幸運可以和你這位獲獎的管理人一同合作。再次恭喜你！

嗯珍妮，再次謝謝妳。我很感謝妳特地打電話來為我祝賀。

Try it! 換你試試看！

WRITING 請依提示寫出完整句子	☐
LISTENING 請聽 MP3 TRACK 38	☐
SPEAKING 請跟著 MP3 唸唸看	☐

1. A I just closed a huge sale this morning.

   B ____________________

   ____

   ____

   (congratulations / big sale)

2. A Hi, Jenny.

   B Oh, hi Steve. ____________________

   ____

   (congratulations / recent wedding)

3. A I'm sorry I was out of the office yesterday. My wife just gave birth to our first child.

   B ____________________

   ____

   (great / hear) (congratulations / become / new father)

4. A Believe me. I was as surprised as anyone that I won the regional sales award.

   B ____________________

   ____

   (deserve) (congratulations / honor)

答案請參閱第 366 頁

UNIT 03

感謝

Expressing Gratitude

I. 表達感謝一定要會的單字片語

MP3 TRACK 39

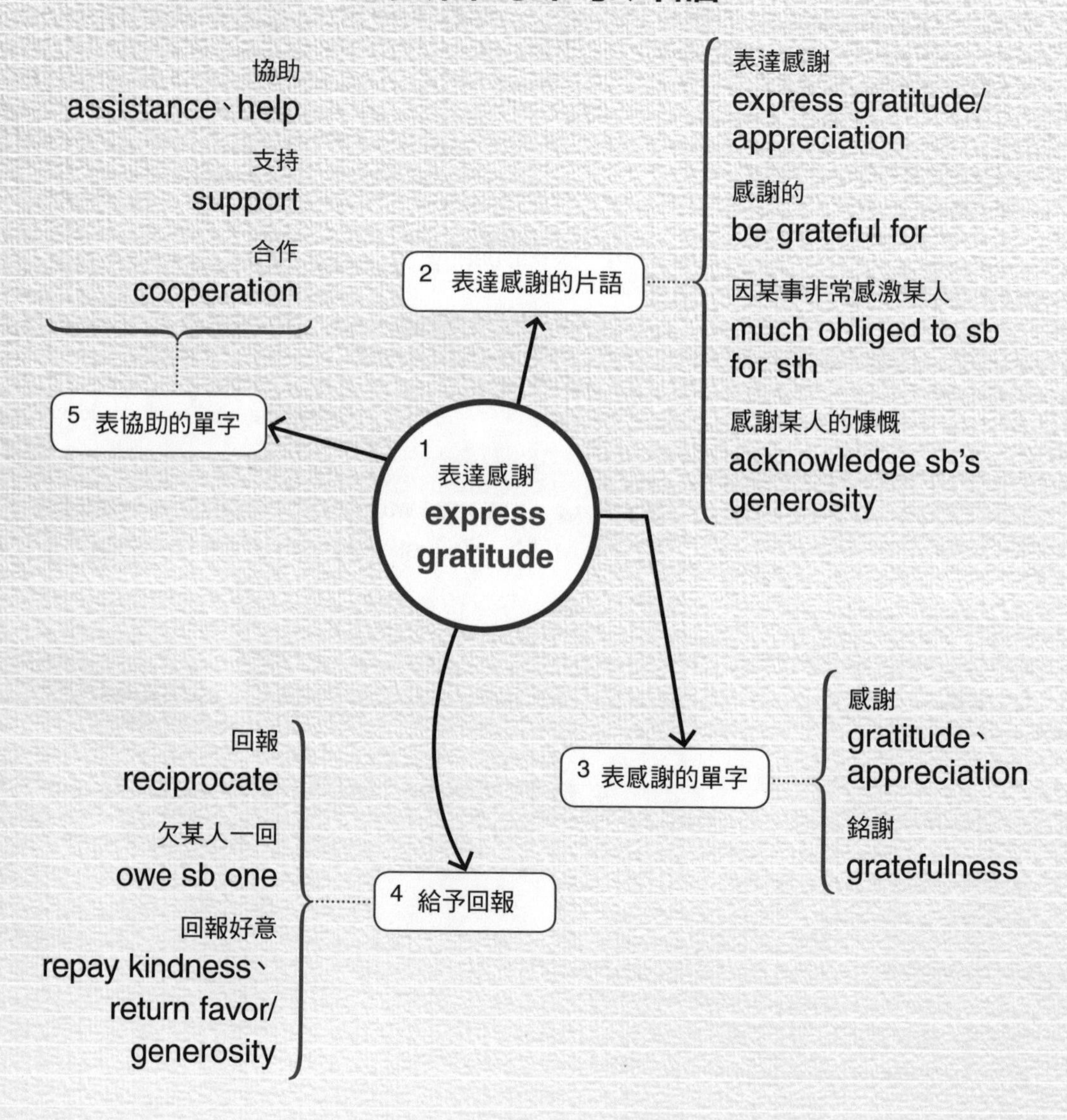

Ⅱ. 表達感謝一定要會的句型

句型 1 表達謝意

I am writing to express my gratitude for sth **with** sth.

例 I am writing to express my gratitude for your help with our company's year-end report.
我寫這封信是想對你在公司年終報告上所給予的協助表達感謝之意。

句型 2 表達給予回報

If you ever need my help in arranging these, please + V.

例 If you ever need my help in arranging these, please tell me.
如果你需要我幫忙安排此事的話，請告訴我。

★ **你也可以這麼說：**

If I can ever assist you in any way, please don't hesitate to let me know.
如果我在任何方面可以幫上忙的話，請不要客氣，儘管讓我知道。

Ⅲ. 用 e-mail 感謝別人的協助與回覆

To: Ken Lee

Subject: Thank You for the Help

Dear Ken,

表示謝意 — I am writing to express my gratitude for your help with our company's year-end report. I really appreciate your assistance with this project. In fact, I don't think I would have been able to finish it by the deadline without your help.

給予回報 — If I can ever assist you in any way, please don't hesitate to let me know. I am always happy to help out.

Once again, thank you.

Sincerely,

Tom

中文翻譯

收件人：李肯恩

主旨：感謝協助

親愛的肯恩：

我寫這封信是想對你在公司年終報告上所給予的協助表達感謝之意。我真的很感謝你在這件案子上的幫忙。事實上，如果沒有你的幫忙，我想我是不可能在截止日期前完成的。

如果在任何方面可以幫上忙的話，請不要客氣，儘管讓我知道。我隨時都很樂意幫忙。

再次感謝你。

謹上

湯姆

To: Tom Yeh
Subject: Re: Thank You for the Help

Dear Tom,

Thank you for the e-mail. Actually, it was my pleasure to work with you on the year-end report. I'm glad to know that my input was of assistance.

All right, Tom. Please let me know if I can ever be of help again.

Yours truly,

Ken Lee

中文翻譯

收件人：葉湯姆
主旨：回覆：感謝協助

親愛的湯姆：

感謝你的來信。事實上，能和你在年終報告共同合作才是我的榮幸。我很高興我的投入有幫上忙。

好啦，湯姆。如果還有機會幫忙的話，務必要讓我知道。

謹上

李肯恩

Ⅳ. 用 e-mail 感謝安排演講事宜及回覆

To: Matt Smith

Subject: Thank You for Arranging Yesterday's Guest Speaker

Dear Matt,

表示謝意

Thank you so much for arranging yesterday's guest speaker. I learned a lot from her presentation at the workshop. I hope to incorporate some of the speaker's tips here in the office.

給予回報

The speakers and workshops that you organize are always very informative. If you ever need my help in arranging these, please tell me. I'd be happy to step up and help.

Thank you again for bringing in such a great guest speaker to our office.

Sincerely,

Katie Lin

中文翻譯

收件人：麥特・史密斯

主旨：感謝安排昨天的演講嘉賓

親愛的麥特：

非常感謝你昨日所安排的演講嘉賓。我從她研討會上的演講獲益良多。我希望可以將她的建議應用在工作上。

你所規劃安排的演講嘉賓及研討會總是非常具有知識性。如果你需要我幫忙安排此事的話，請告訴我。我會很高興站出來幫忙的。

再次感謝你介紹公司這麼棒的演講嘉賓。

謹上

林凱蒂

To: Katie Lin

Subject: Re: Thank You for Arranging Yesterday's Guest Speaker

Dear Katie,

Thank you for your kind e-mail. I was glad to hear that you enjoyed yesterday's guest speaker so much.

I enjoy organizing the speakers for our company's workshops, and it makes it all worthwhile to know that people in our company are benefiting from these events. If I ever need your help in the future with them, I will be sure to let you know.

Thank you again for the message, Katie. I look forward to seeing you at our next workshop.

Sincerely,

Matt Smith

中文翻譯

收件人：林凱蒂

主旨：回覆：感謝安排昨天的演講嘉賓

親愛的凱蒂：

感謝妳友善的來信。我很高興聽到妳對於昨日的演講感到非常滿意。

我很樂意為公司的研討會安排演講嘉賓，知道公司同仁從這些活動中獲益就讓我覺得一切都是值得的。如果未來有需要妳幫忙的地方，我一定會讓妳知道的。

凱蒂，再次感謝妳的訊息。我期待在下次的研討會上見到妳。

謹上

麥特·史密斯

V. 用 e-mail 表達感謝贈禮與回覆

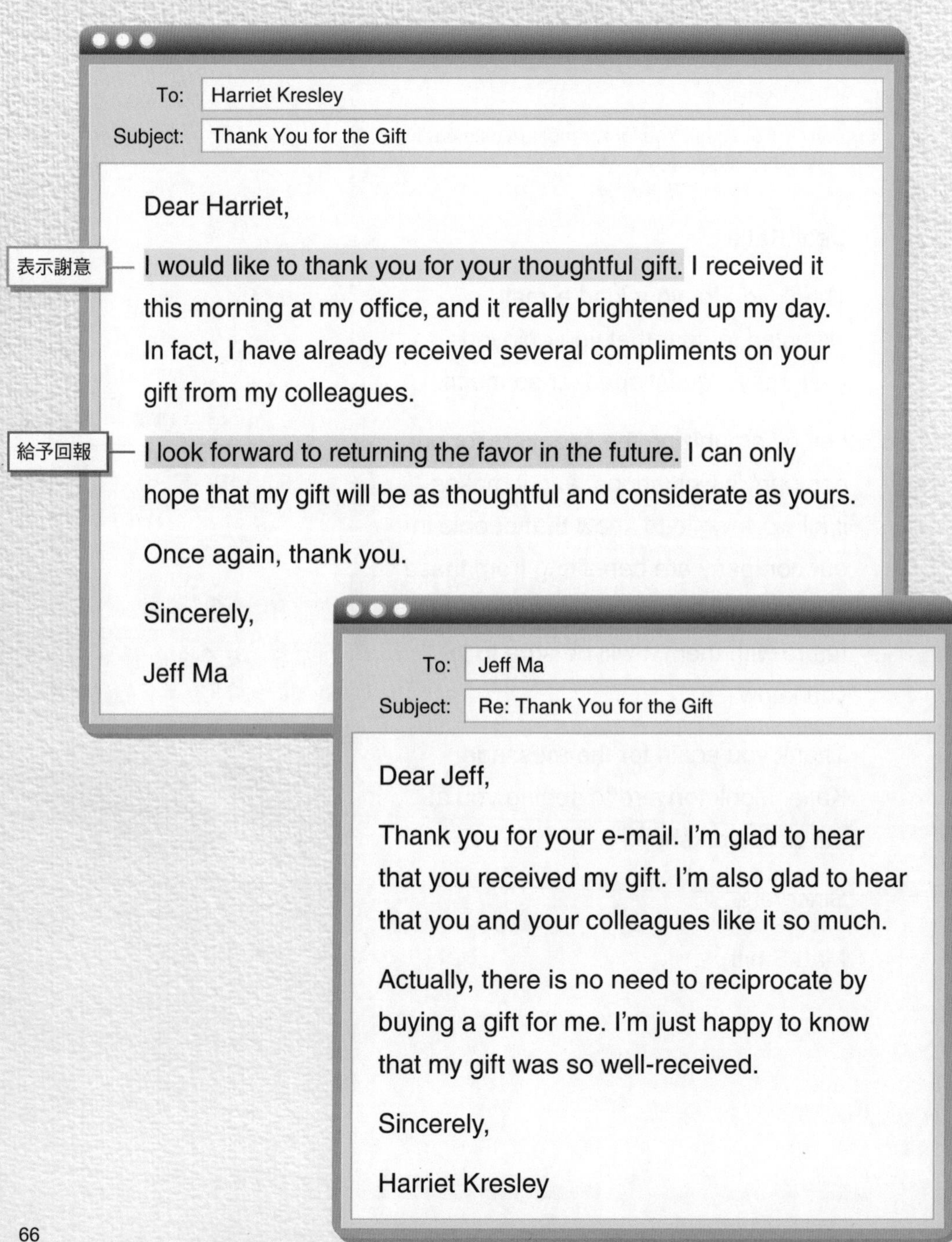
To: Harriet Kresley
Subject: Thank You for the Gift

Dear Harriet,

表示謝意 — I would like to thank you for your thoughtful gift. I received it this morning at my office, and it really brightened up my day. In fact, I have already received several compliments on your gift from my colleagues.

給予回報 — I look forward to returning the favor in the future. I can only hope that my gift will be as thoughtful and considerate as yours.

Once again, thank you.

Sincerely,

Jeff Ma

To: Jeff Ma
Subject: Re: Thank You for the Gift

Dear Jeff,

Thank you for your e-mail. I'm glad to hear that you received my gift. I'm also glad to hear that you and your colleagues like it so much.

Actually, there is no need to reciprocate by buying a gift for me. I'm just happy to know that my gift was so well-received.

Sincerely,

Harriet Kresley

中文翻譯

收件人：哈利葉．克雷斯利
主旨：感謝贈禮

親愛的哈利葉：

我是要寫信謝謝你貼心的禮物。我今天早上在辦公室收到了禮物，讓我整天心情都很好。事實上，我很多同事對你的這份禮物讚譽有加。

我期待將來能夠回報你。我只希望我的禮物可以像你送我的一樣體貼用心。

再次感謝你。

謹上

馬傑夫

收件人：馬傑夫
主旨：回覆：感謝贈禮

親愛的傑夫：

謝謝你的來信。我很高興聽到你收到了我的禮物。我也很高興聽到你和你同事都很喜歡。

事實上，不需要買禮物來回贈給我。聽到我的禮物受到歡迎就讓我很開心了。

謹上

哈利葉．克雷斯利

Try it! 換你試試看!

1. 我寫信是要感謝你的幫忙。

2. 感謝你貼心的禮物。

3. 如果在任何方面可以幫上忙的話，請不要客氣，儘管讓我知道。

4. 我期待將來能回報你的好意。

答案請參閱第 366 頁

VI. 用電話表達感謝一定要會這樣說

MP3 TRACK 40

感謝業務上的協助

A **I'm calling to say thanks for your help on** the Patterson case.

我打電話來是要謝謝你對派德森這件案子的協助。

B Don't mention it. I'm glad that everything went well.

不用客氣。我很高興事情都進行得很順利。

表示不用客氣的說法還有：
sure、sure thing、it's all right、it's nothing 等。

A **Thanks a million for your assistance** this morning.

非常感謝你今天早上的協助。

B No problem. I'm always happy to step in and lend a hand.

不用客氣。我隨時都願意參與幫忙。

感謝贈禮

A **I just called to say how much I appreciate your gift.**

我只是要打電話跟你說我很感謝你的禮物。

B I'm glad you like it. I thought it would be perfect for you.

很高興你喜歡。我就覺得它很適合你。

A **Thanks so much for the thoughtful gift you sent to me** this morning.

非常謝謝你今天早上送我這樣窩心的禮物。

B I'm glad to hear that you got it and that you like it.

我很高興聽到你收到禮物而且喜歡。

感謝演講安排

A **I just wanted to let you know how much I enjoyed the speech** you gave yesterday.

我只是想要讓你知道我有多喜歡你昨天的演講。

B Thanks. I hope that you picked up some information that will be of benefit to you.

謝謝。我希望你能從中獲得一些有用的資訊。

A The guest speaker was very informative. **Thank you for setting this up.**

這位演講人提供很多情報。謝謝你的安排。

B My pleasure. I'm glad that you enjoyed the speech.

這是我的榮幸。我很高興你喜歡這場演講。

Ⅶ. 用電話表達感謝

史坦打電話給蘇表示感謝，因為有她的幫忙才讓案子準時完成。

LISTENING 請聽 MP3 TRACK 41 ☐ | SPEAKING 請跟著 MP3 唸唸看 ☐

Stan: Hi. This is Stan Hu at ComCorp. Is Sue Gordon there, please?

Sue: This is Sue. How are things, Stan?

Stan: Great, now that my project is finished. I want to thank you again for all of your assistance with this.

表示謝意

Sue: Oh, that's all right. I'm just glad that I could give you a hand.

Stan: You did more than that. You saved the day.

Sue: I wouldn't go that far, Stan.

Stan: No, it's true. I was really in a crunch, and I don't think that I would have been able to get that project in on time without your help.

Sue: You were on the right track. I just came in and helped speed things up a bit.

Stan: Well, I owe you one. So if you ever need any assistance with anything in the future, please don't hesitate to let me know.

給予回報

Sue: Thanks. I'll be sure to let you know.

上班族加油站

save the day/situation 是指成功避免事情走向尷尬、不愉快、失敗的處境，表示「化險為夷；轉敗為勝」。

嗨。我是聯合企業的胡史坦。請問蘇·戈爾登在嗎?

我是蘇。史坦，最近怎麼樣?

很好，現在我的案子完成了。我要再次感謝妳在這件事上所提供的一切協助。

喔，不用客氣。能幫上你的忙就讓我很高興了。

妳做的不只這些，妳讓我化險為夷。

史坦，我不覺得有到那個地步。

不，這都是事實。我那時真的是處境艱難，如果沒有妳的幫忙，我想我是沒有辦法準時完成這件案子的。

你的方向是正確的。我只是參與並協助讓事情進行得快一點而已。

嗯，我欠妳一個人情，所以如果妳日後有需要任何協助的話，不要遲疑，務必要讓我知道。

謝謝。我一定會讓你知道的。

Try it! 換你試試看!

WRITING 請依提示寫出完整句子	☐
LISTENING 請聽 MP3 TRACK 42	☐
SPEAKING 請跟著 MP3 唸唸看	☐

1. Ⓐ Did you get that report done on time?

   Ⓑ ______________________________

   ____

   (yes / express / gratitude / help)

2. Ⓐ Happy birthday, Kathy.

   Ⓑ Thanks, Tom. ______________________________

   ____

   (thank / thoughtful / gift)

   Ⓐ I'm glad you like it.

3. Ⓐ I'm glad that I was able to help you out at the meeting yesterday.

   Ⓑ ______________________________

   ____

   (ever / assist / way / not / hesitate / know)

4. Ⓐ I sent in the regional sales report a week before the deadline. I hope that this was helpful to you.

   Ⓑ ______________________________

   ____

   (yes / look forward / return / favor / future)

答案請參閱第 366 頁

UNIT 04 表達慰問

Expressing Concern

I. 表達慰問一定要會的單字片語

MP3 TRACK 43

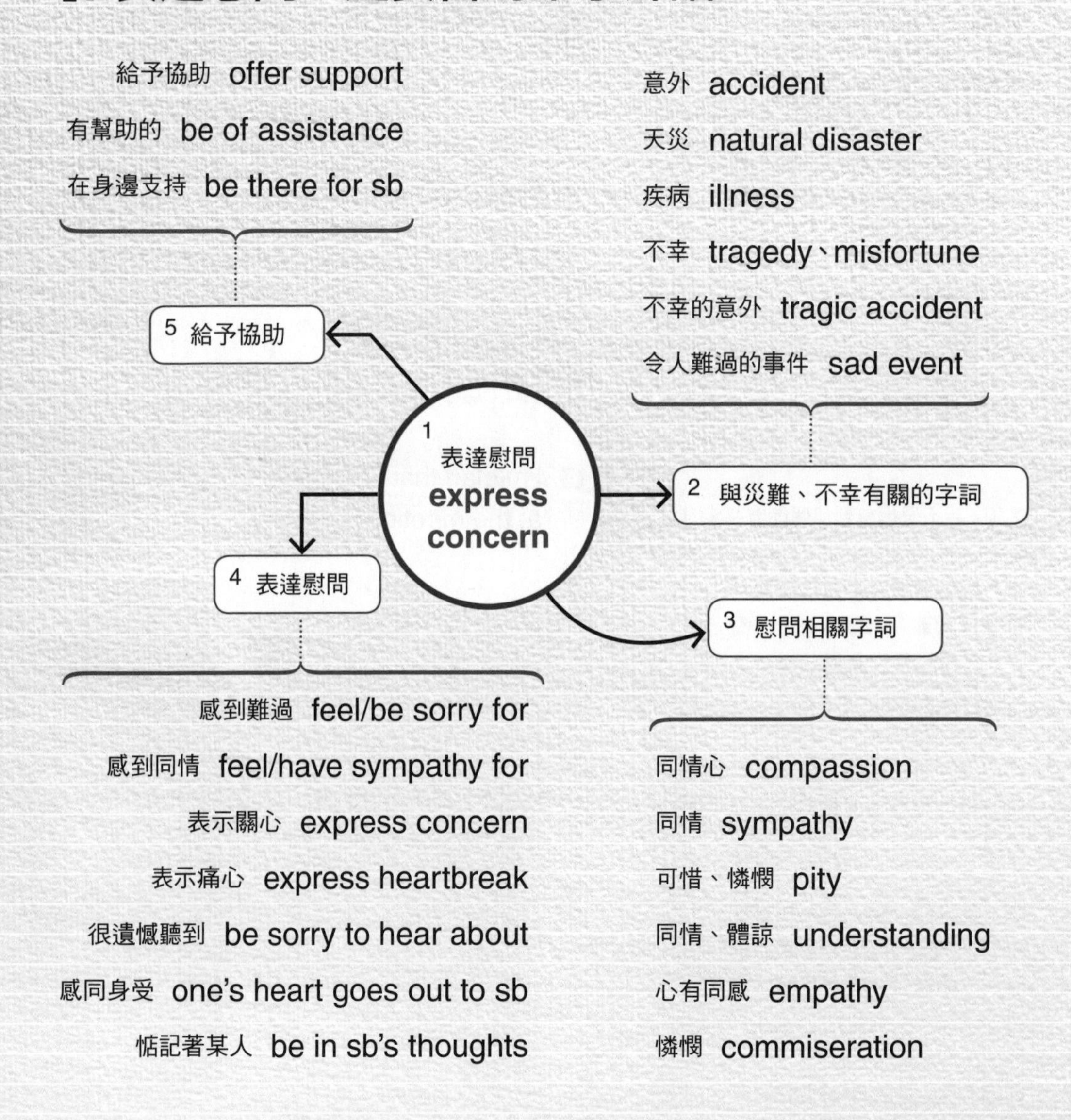

Ⅱ. 表達慰問一定要會的句型

句型 1 得知壞消息

I am sorry to hear about sth.

例 I am sorry to hear about this horrible disaster.
我很遺憾聽到這麼可怕的災難。

句型 2 表達同情

My deepest sympathies go out to sb.

例 My deepest sympathies go out to you and the people of your country at this time.
在這個時刻，我對你以及你的國人同胞表達最深切的同情。

句型 3 提供協助

If there's anything I can do for you . . .

例 If there's anything I can do for you while you are recovering, please let me know.
在你康復這段期間如果有什麼我可以做的，請讓我知道。

句型 4 給予祝福

Our prayers are with sb.

例 Our prayers are with you.
我們會替你禱告。

Ⅲ. 如何用 e-mail 表達對地震的慰問與回覆

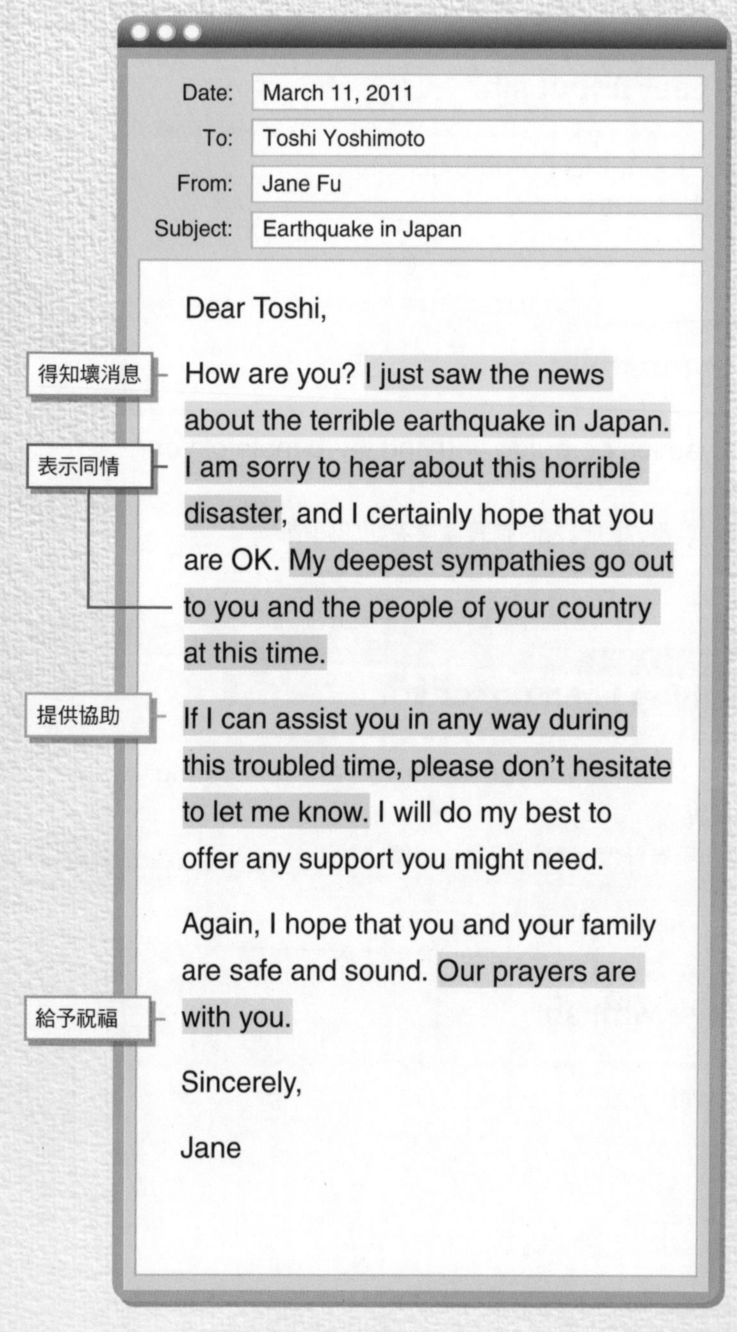

中文翻譯

日期：二〇一一年三月十一日
收件人：俊．吉本
寄件人：傅珍
主旨：日本地震

親愛的俊：

你好嗎？我剛看到一則關於日本發生可怕地震的新聞。我很遺憾聽到這麼可怕的災難，當然希望你沒事。在此時刻，我對你及你的國人同胞表達最深切的同情。

在這個混亂不安的時候，如果在任何方面我幫得上忙的話，請不要客氣，儘管讓我知道。我會盡我所能提供可能需要的協助。

再次希望你和你的家人安然無恙。我們會替你禱告。

謹上

珍

Date:	March 13, 2011
To:	Jane Fu
From:	Toshi Yoshimoto
Subject:	Re: Earthquake in Japan

Dear Jane,

Thank you for your e-mail. We are safe here in Kyoto. The earthquake's epicenter was actually in the northern part of Japan, so we weren't affected by this quake that much.

Even though we are fine, our hearts do indeed go out to all the others in Japan who have been affected by this quake.

Thank you again for your e-mail. It was comforting to hear from you during this difficult time.

Yours truly,

Toshi

上班族加油站

其他與地震相關的詞彙還有 the Richter scale（芮氏規模）、magnitude（震度）、tsunami（海嘯）等。

中文翻譯

日期：二〇一一年三月十三日
收件人：傅珍
寄件人：俊・吉本
主旨：回覆：日本地震

親愛的珍：

感謝妳的來信。我們京都這裡很安全。震央其實是在日本北部，所以這次地震對我們來說影響並沒有那麼大。

儘管我們沒事，我們確實心繫著日本其他地方受此次地震影響的人。

再次感謝妳的來信。在這個艱困的時刻聽到妳的消息令人很欣慰。

謹上

俊

Ⅳ. 如何用 e-mail 表達對車禍的慰問及回覆

To: Molly Brown

Subject: Are you alright?

Dear Molly,

得知壞消息

I am writing this e-mail to make sure that you are all right after your recent car accident. I was shocked to learn this morning that you had been in a car crash a few days ago.

表示同情、感同身受

My heart goes out to you. I know that the accident must have been a frightening experience.

提供協助

Please remember that I am always here for you. Should you need any support in any way, just let me know.

給予祝福

I look forward to getting an update from you soon. You are in my thoughts and prayers.

Sincerely,

Ken Lo

上班族加油站

you are in my thoughts 字面是表示「我會時常惦記著你」，常用於對方不如意時，希望藉此讓對方心裡感到慰藉。

中文翻譯

收件人：茉莉・布朗
主旨：妳還好嗎？

親愛的茉莉：

我寫這封信是要確認妳前陣子車禍後是否無恙。我今天早上聽到妳幾天前發生車禍感到非常震驚。

我對妳的遭遇感同身受。我知道車禍一定是非常可怕的經驗。

請記得我隨時都在妳身邊。如果有需要任何支援的地方，就讓我知道。

我希望能盡快得知妳的近況。我會掛念著妳並為妳禱告。

謹上

羅肯恩

To: Ken Lo

Subject: Re: Are You Alright?

Dear Ken,

Thank you for the e-mail this morning. Please don't worry—I am OK. I'm still a little shaken up[1] after the accident, but I was not injured. My car, however, was totaled.[2]

It's been a difficult time, and during times like these, it is nice to know that you can count on the support of your friends. Thanks again for writing me this morning.

Take care, and stay safe on the roads.

Best regards,

Molly

Vocabulary & Phrases

1. shake up [ʃek] *v.*
   嚇到某人（常用被動語態來表示）
2. total [ˋtotl̩] *v.*
   完全毀壞（口語用法）

中文翻譯

收件人：羅肯恩
主旨：回覆：妳還好嗎？

親愛的肯恩：

感謝你今天早上的來信。請不用擔心——我沒事。在意外過後我還是有點驚魂未定，不過我並沒有受傷。只是我的車子全毀了。

這段時間很難熬，在這樣的時刻，很高興知道有你這樣的朋友的支持可以依靠。再次謝謝你今天早上的來信。

保重，路上要小心。

謹上

茉莉

V. 如何用 e-mail 表達對手術的慰問及回覆

上班族加油站

go to the hospital 表示「去醫院看病」、be in the hospital 表示「住院中」、be out of hospital 則表示「出院」。

To: Mavis Saples

Subject: Get Well Soon

Dear Mavis,

得知壞消息 — I am writing to inquire about your health after your recent surgery. My colleague, Tony Nie, mentioned that you went in to the hospital yesterday for this surgery.

表示同情 — I was sorry to hear about your recent health issue. I certainly hope that the operation was a success and that you are feeling better now.

提供協助 — If there's anything I can do for you while you are recovering, please let me know.

給予祝福 — All the best to you, and I wish you a speedy recovery. We will be thinking about you.

Warm regards,

Greg

To: Greg Chu

Subject: Re: Get Well Soon

Dear Greg,

Thank you for your considerate e-mail. I did indeed have surgery the other day, but I am feeling fine now. In fact, I'm feeling better every day.

It was nice to hear from you during my recovery. Your e-mail certainly put a smile on my face. Thanks.

Sincerely,

Mavis Saples

中文翻譯

收件人：瑪維絲．薩波絲
主旨：早日康復

親愛的瑪維絲：

我寫信是要詢問妳最近手術後的健康狀況。我的同事聶東尼提到妳昨天住院動手術。

我很遺憾聽到妳最近的健康問題。我非常希望妳手術成功，並且漸有起色。

在妳康復這段期間如果有什麼我可以做的，請讓我知道。希望妳一切順利，早日康復。我們會一直掛念妳的。

請多保重

葛瑞格

收件人：楚葛瑞格
主旨：回覆：早日康復

親愛的葛瑞格：

感謝你體貼的來信。我幾天前的確動了手術，但是現在沒問題了。事實上，我每天都覺得越來越好。

在康復期間很高興聽到你的消息。你的來信的確讓我很開心。謝謝你。

謹上

瑪維絲．薩波絲

Try it! 換你試試看!

1. 我無法告訴你得知你最近的不幸消息我有多傷心。

(得知 learn about)

2. 我希望你現在有覺得好一點，並且盡快康復。

(康復 recover)

3. 請瞭解如果你需要任何幫助的話，你有一個可以依賴的朋友。

(依賴 rely on)

4. 在這難熬的時刻，我們會一直惦念著你。

答案請參閱第 367 頁

VI. 用電話表示慰問一定要會這樣說

MP3 TRACK 44

得知壞消息

A: **Did you hear about** the terrible accident on the highway nearby?

你有聽說附近高速公路上發生的嚴重車禍嗎？

B: Yes, I just heard about it on the radio.

有，我剛從廣播上聽到。

A: **I just saw on** Facebook **that** a hurricane hit Florida.

我剛從臉書上看到有颶風侵襲佛羅里達。

B: Yes, you're right. Thankfully, we are all OK here.

是啊，沒錯。謝天謝地，我們這裡一切都很平安。

表達同情

A: Our boss just found out some bad news. He has cancer.

我們老闆剛得知噩耗。他得了癌症。

B: **My heart goes out to** him and his family **at this bad moment**.

在這讓人難過的時候，我與他和他的家人感同身受。

提供協助

A **If I can do anything for you** during your recovery, **be sure to let me know.**

在你復元期間如果有什麼我可以做的事，請一定要讓我知道。

B Thanks. I'm sure I'll be up and on my feet again soon.

謝謝。我相信我很快就會再振作、重新站起來。

給予祝福

A **If it's any comfort, we are praying for you** at work.

如果能給你安慰，我們會在工作時為你祈禱的。

B Thanks. That means a lot to me.

謝謝。那對我來說意義重大。

A I have a long and difficult road ahead of me.

在我前方是條又長又艱困的路。

B **We'll be with you in spirit** every step of the way.

未來每一步我們的精神都與你同在。

Ⅶ. 如何用電話表示慰問

喬治・卡爾打電話給工作夥伴朵蒂・班德告知同事出車禍的消息。

LISTENING 請聽 MP3 TRACK 45 ☐	**SPEAKING** 請跟著 MP3 唸唸看 ☐

George: Hi. I'm looking for Dottie Bender, please.

Dottie: Speaking. How can I help you?

George: This is George Karl at Excel Enterprises. Unfortunately, I have some bad news. Todd Pampalone was involved in a car accident this morning.

告知壞消息

Dottie: Oh, my God! Is he all right?

George: Well, he's in the hospital right now. He'll probably be there for the next days as well.

表示同情

Dottie: I'm so sorry to hear this. I was just talking with him on the phone yesterday.

George: Yes, I know that he is your primary contact here at Excel. That's why I wanted to let you know about this right away.

Dottie: Thank you, George. I would have been wondering why Todd hadn't be in contact with me.

提供協助

George: Well, if you require any assistance while Todd is recovering, please let me know. I'll be glad to fill in for Todd while he is in the hospital.

給予祝福

Dottie: Thank you. And please wish Todd a speedy recovery, if you see him.

嗨，麻煩找朵蒂·班德。

我就是。我可以怎麼幫你？

我是伊克塞爾企業的喬治·卡爾。不幸地，我有一些壞消息。陶德·潘帕隆今早出了車禍。

喔，天啊！他還好嗎？

嗯，他現在在醫院裡。可能接下來幾天還會待在醫院。

我很遺憾聽到這件事。我昨天才和他通過電話的。

是，我知道他是你們伊克塞爾的主要聯絡窗口。這也是為什麼我要立即通知你們。

謝謝你，喬治。我可能還納悶陶德為什麼一直沒和我聯絡呢。

那麼，陶德康復這段期間如果你們需要任何協助的話，請讓我知道。陶德住院時我很樂意暫代他的工作。

謝謝。如果你見到陶德的話，請祝他早日康復。

Try it! 換你試試看!

WRITING	請依提示寫出完整句子	☐
LISTENING	請聽 MP3 TRACK 46	☐
SPEAKING	請跟著 MP3 唸唸看	☐

1. Ⓐ Did you hear about the earthquake in Asia?

   Ⓑ ______________________________

   ____

   (yes / see / news / earthquake / TV)

2. Ⓐ It's nice of you to come visit me at the hospital.

   Ⓑ ______________________________

   ____

   (sorry / hear / health issue)

3. Ⓐ It has been a difficult moment for us.

   Ⓑ ______________________________

   ____

   (remember / always / here / you)

4. Ⓐ I start chemotherapy treatments for my cancer next week.

   Ⓑ ______________________________

   ____

   (prayers / you)

答案請參閱第 367 頁

UNIT 05 弔唁
Offering Condolences

I. 弔唁一定要會的單字片語

MP3 TRACK 47

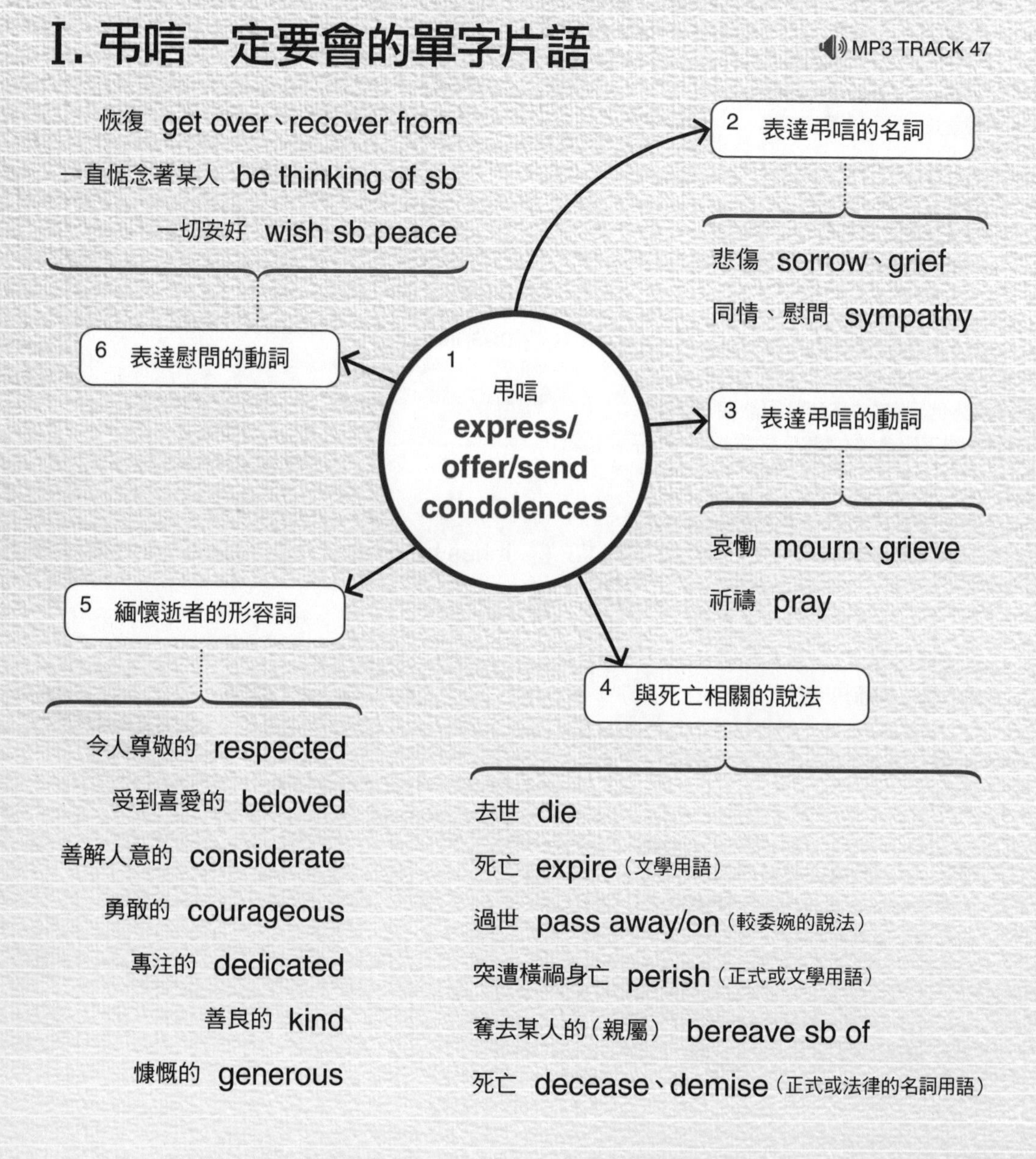

Ⅱ. 弔唁一定要會的句型

句型 1 表達得知消息時感到哀傷

It was with great sorrow that S. + V.

例 It was with great sorrow that I learned of the death of your husband.
得知你丈夫過世的消息讓我感到相當哀傷。

句型 2 表示哀悼

Please allow me to express sth.

例 Please allow me to express my deepest sympathy to you at this time.
請容我在這個時刻向你表達最深切的同情。

句型 3 向逝者致敬

His passing will leave sth.

例 His passing will leave a void in the hearts of those who knew him.
他的離去一定會讓認識他的人感到悵然。

句型 4 表達慰問之意

You are in my thoughts during + 時間

例 You and your family are in my thoughts during this moment of grief.
在這段哀傷的日子我會惦念著你和你的家人。

Ⅲ. 如何用 e-mail 寫弔唁信

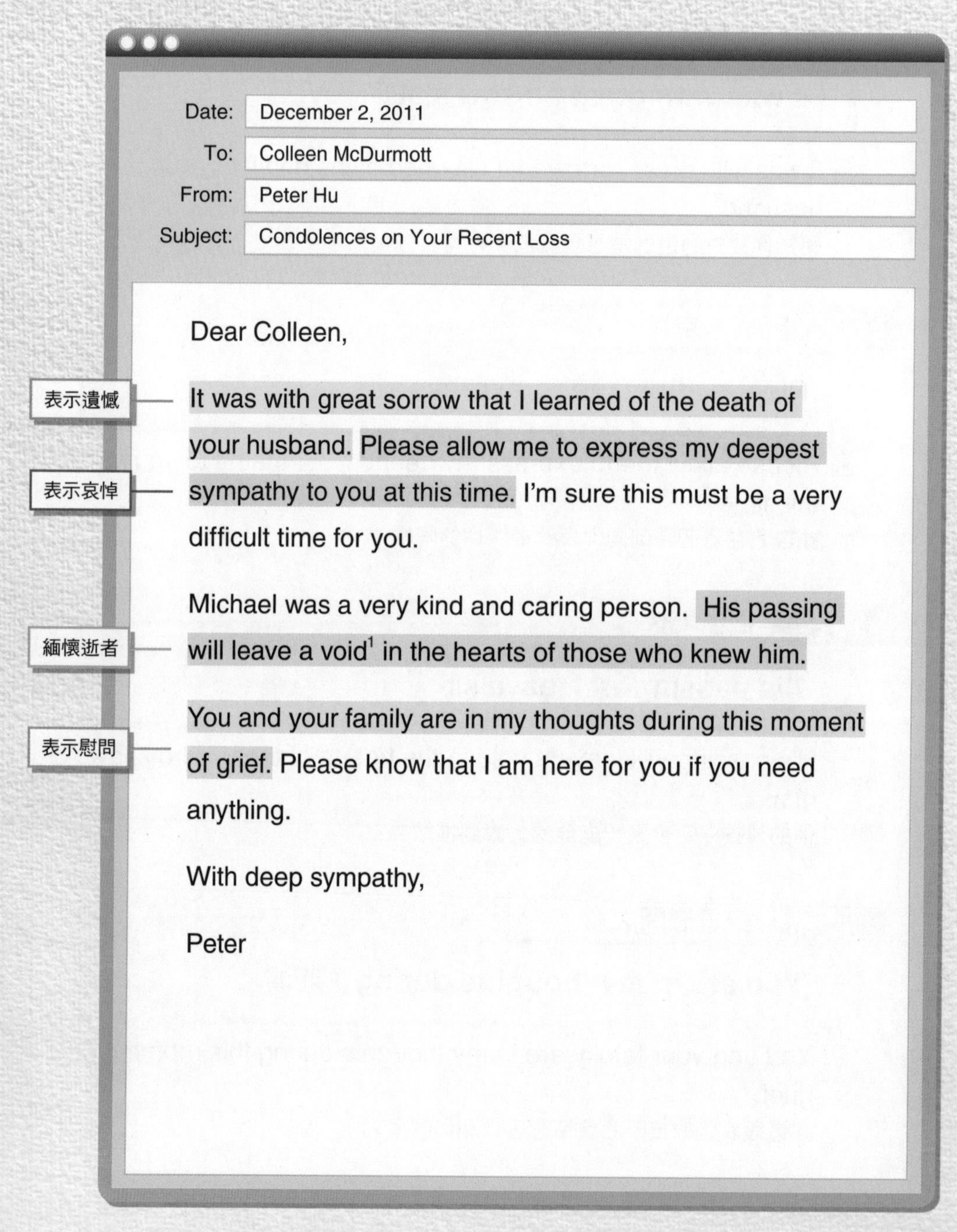

Date: December 2, 2011

To: Colleen McDurmott

From: Peter Hu

Subject: Condolences on Your Recent Loss

Dear Colleen,

It was with great sorrow that I learned of the death of your husband. Please allow me to express my deepest sympathy to you at this time. I'm sure this must be a very difficult time for you.

Michael was a very kind and caring person. His passing will leave a void[1] in the hearts of those who knew him.

You and your family are in my thoughts during this moment of grief. Please know that I am here for you if you need anything.

With deep sympathy,

Peter

Vocabulary & Phrases

1. void [vɔɪd] *n.* 空虛；空虛感

中文翻譯

日期：二〇一一年十二月二日
收件人：珂琳．麥克德莫
寄件人：胡彼特
主旨：弔慰失去摯愛

親愛的珂琳：

得知妳丈夫過世的消息讓我感到十分哀傷。請容我在這個時刻向妳表達最深切的同情。我相信這段時間妳一定很不好受。

麥克是一個非常善良又關心別人的人。他的離去一定會讓認識他的人感到悵然。

在這段哀傷的日子我會惦念著妳和妳的家人。請記得如果有任何需要的話，我都會在妳身邊。

獻上深切同情

彼特

文化補給站

參加西方喪禮要不要包禮金？

在西方要不要參加喪禮是根據你與亡者（the deceased）與喪家（the bereaved）之間的親疏關係而定，並沒有一定的禮儀或法則。如果你覺得彼此的關係並不是那麼親近，則可以選擇寄送花籃、並附上親自書寫的弔唁信來表達慰問。

西方文化中並沒有白包的習俗，有時候喪家也會謝絕送花，而是請求送葬者（mourner）以亡者之名捐贈款項給慈善機構或公益組織。要留意的是，這種捐贈要在喪禮（funeral service）前七到十天之內完成，這樣喪家才能將捐贈名冊公布於喪禮上。而進行捐贈時，捐贈人要確認受贈機構知道喪家名諱及住址，這樣才能告知喪家有這筆捐贈，而捐贈者可以在弔唁信上表明有此捐贈，但無須提到捐贈金額。

Ⅳ. 如何用 e-mail 回覆弔唁信

Date:	December 5, 2011
To:	Peter Hu
From:	Colleen McDurmott
Subject:	Re: Condolences on Your Recent Loss

Dear Peter,

Thank you for your e-mail. Your sympathy is appreciated.

Your words of support have been helpful in[1] getting me through[2] this difficult time.

Thank you again for your e-mail, Peter. It was good to hear from you.

Yours truly,

Colleen

Vocabulary & Phrases

1. be helpful in [ˋhɛlpfəl] *phr.* 在……有幫助的
2. get through [θru] *v.* 度過難關

中文翻譯

日期：二〇一一年十二月五日
收件人：胡彼特
寄件人：珂琳．麥克德莫
主旨：回覆：弔慰失去摯愛

親愛的彼特：

感謝你的來信。我很感激你的慰問。

你支持的話語在我度過這個難關時幫了我很多忙。

彼特，再次感謝你的來信。很高興聽到你的消息。

謹上

珂琳

Try it! 換你試試看!

1. 我們很遺憾聽到你蒙受巨大的損失。

2. 請接受我誠摯的哀悼之意。

3. John 讓每個認識他的人都難過地懷念著他。

4. 我們在祈禱時會記得你和你的家人。

答案請參閱第 367 頁

V. 電話弔唁一定要會這樣說

MP3 TRACK 48

得知逝世消息

A **I was sorry to hear about the recent death** in your family.
我很遺憾聽到最近你家人過世的消息。

B Thank you. It has been a difficult time for all of us.
謝謝。對我們家人而言這段日子很不好受。

A **I am grief-stricken over the death of** your husband.
對於妳丈夫的逝世我感到極為悲痛。

B Yes, it was totally unexpected.
是啊，這完全出乎意料之外。

表示哀悼

A **Please accept our condolences on** this sad event.
請接受我們對此悲傷事件的哀悼之意。

B We certainly appreciate it.
我們相當感激。

A **You have our sincere condolences.**
我們向你表達誠摯的慰問之意。

B Thank you very much.
非常感謝你們。

對逝者致敬

A Candice **will be sadly missed by all**. Her presence brought a smile to the face of everyone who knew her.

坎蒂絲將讓我們所有人難過地想念她。她在的時候為每一位認識她的人帶來歡樂。

B Thank you. I appreciate your kind words.

謝謝你。我很感激你善意的話語。

給予協助

A **If there's any way I can help you through this difficult moment, just tell me.**

如果有可能幫上忙讓你熬過這段日子的地方，儘管告訴我。

B Your thoughts mean a lot to me.

你的關懷對我來說很重要。

結尾慰問

A **I wish you peace as you grieve the loss of your loved one.**

我希望你在為失去所愛而感到哀傷時一切都安好。

B Thank you.

謝謝你。

VI. 如何用電話弔唁

弗萊德得知工作夥伴海倫失去至親的消息，打電話向她表達慰問之意。

LISTENING 請聽 MP3 TRACK 49 ☐ SPEAKING 請跟著 MP3 唸唸看 ☐

Fred: Hello. Is Helen there, please?

Helen: Yes, this is Helen.

Fred: Helen, this is Fred over at Amcro. I'm calling because I just learned of the recent loss in your family.

表示遺憾

Helen: Oh, yes. It was very unexpected.

Fred: Well, I was sad to hear this news. You have my complete sympathy.

表示哀悼

Helen: It has been a difficult time for us.

Fred: If there is anything I can do for you at this time, please don't hesitate to let me know.

Helen: Thank you. I think I'll be able to manage.

表示慰問

Fred: Well, I wish you peace as you grieve over your loss.

Helen: Thank you, Fred. That means a lot.

你好。請問海倫在嗎？

是，我就是海倫。

海倫，我是艾莫科的弗萊德。我打來是因為聽到最近妳家人過世的消息。

喔，是啊。這件事出乎意料之外。

嗯，我很難過聽到這個消息，對此我深表同情。

發生這個事情以來，我們都很難過。

這段時間如果有什麼我可以幫得上忙的地方，請讓我知道。

謝謝。我想我可以應付得來。

那麼，我希望妳在哀悼失去所愛時一切安好。

謝謝你，弗萊德。那對我來說很重要。

Try it! 換你試試看!

WRITING 請依提示寫出完整句子	☐
LISTENING 請聽 MP3 TRACK 50	☐
SPEAKING 請跟著 MP3 唸唸看	☐

1. Ⓐ Did you hear the news about the death in Greg's family?

   Ⓑ ______________________________

   ____

   (yes / saddened / learn / loss / family)

2. Ⓐ I just found out that my aunt passed away.

   Ⓑ ______________________________

   ____

   (allow / express / deepest sympathy / time)

3. Ⓐ Did you hear? Humphrey passed away last week.

   Ⓑ ______________________________

   ____

   (sorry / hear / his passing)

4. Ⓐ Thanks for your call.

   Ⓑ ______________________________

   ____

   (here / you / need / anything)

答案請參閱第 367 頁

UNIT 06 | 邀請 Invitations

I. 邀請一定要會的單字片語

MP3 TRACK 51

Ⅱ. 邀請一定要會的句型

句型1 活動時間地點

You are cordially invited to + 活動 + on 日期 + at 時間 + at 地點

例 You are cordially invited to my son's upcoming wedding on February 6, 2012, at 2 p.m. at the Northside Methodist Church.
誠摯地邀請你參加小犬的婚禮，時間、地點謹訂於二〇一二年二月六日下午兩點於北方衛理教堂舉行。

句型2 服裝或其他規定

Please note that S. + V.

例 Please note that the dress code is semiformal.
請留意需穿著半正式的服裝。

句型3 交通位置

S. is located at/in + 地方

例 Northside Methodist Church is located at the corner of State Street and Fallstaff Drive.
北方衛理教堂位於史塔特街與弗斯塔夫大道轉角處。

句型4 竭誠歡迎蒞臨

We look forward to + V-ing.

例 We look forward to seeing you on this special day.
我們期盼在這個特別的日子能見到你。

Ⅲ. 如何用 e-mail 邀請參加婚禮與回覆

To: Tom Peterson

Subject: Wedding Invitation

Dear Tom,

You are cordially invited to my son's upcoming wedding on February 6, 2012, at 2 p.m. at the Northside Methodist Church.

We would be very happy if you could take part in this joyous event, and we extend this invitation also to your family.

Please note that the dress code is semiformal.

Northside Methodist Church is located at the corner of State Street and Fallstaff Drive. Parking is available nearby. A map and detailed directions and can be found at this link: Northside Methodist Church.

You can contact me by phone or e-mail to confirm your attendance. We look forward to seeing you on this special day.

Sincerely,

Helen

中文翻譯

收件人：湯姆．彼特森
主旨：婚禮邀請

親愛的湯姆：

誠摯地邀請你參加小犬的婚禮，時間、地點謹訂於二〇一二年二月六日下午兩點於北方衛理教堂舉行。

若你能來參加這個喜事我們會感到非常高興，我們也同樣邀請你的家人參加。

請留意需穿著半正式的服裝。

北方衛理教堂位於史塔特街與弗斯塔夫大道轉角處。附近有停車位。地圖和詳細的方位請見此連結：Northside Methodist Church。

你可以用電話或電子郵件跟我聯絡確認是否出席。我們期盼在這個特別的日子能見到你。

謹上

海倫

To: Helen Lu
Subject: Re: Wedding Invitation

Dear Helen,

Thank you for your e-mail. Congratulations on your son's upcoming wedding!

My wife and I would be delighted to attend.

Thank you once again.

Yours truly,

Tom

中文翻譯

收件人：盧海倫
主旨：回覆：婚禮邀請

親愛的海倫：

感謝妳的來信。恭喜妳兒子即將舉行婚禮！

我太太和我很樂意參加。

再次感謝妳。

謹上

湯姆

Ⅳ. 如何用 e-mail 寫聖誕派對邀請函與回覆

To: Ken Kersey
Subject: Christmas Party Invitation

Dear Ken,

活動時間地點 — Amtex will be hosting a Christmas party on December 23 at Tower Hotel, and we would be honored if you could come and join us. Company staff, along with many of our valued customers and suppliers, will be attending.

Drinks and appetizers will be served at 6 p.m., and a sit-down dinner will begin at 7 p.m.

交通位置 — The Tower Hotel is located at 246 Seaside Avenue. For your convenience, here is a link to a map with the exact address along with detailed directions: Tower Hotel.

表示歡迎 — We certainly hope you can make it. Please RSVP when you get a chance.

Sincerely,

Paul

上班族加油站

RSVP 為法語 repondez s'il vous plait 的縮寫，常用在邀請函中，表示希望能得到回覆的意思。

中文翻譯

收件人：肯恩・柯爾西
主旨：聖誕派對邀請

親愛的肯恩：

艾姆泰克斯將於十二月二十三日在淘兒飯店舉辦聖誕派對，如果你能來共襄盛舉將是我們的榮幸。公司員工以及我們許多重要客戶和供應商都將會參加。

飲料及開胃點心會在晚上六點供應，正式晚宴將於七點開始。

淘兒飯店位於海濱路二四六號。為了你的方便，這裡附上地圖連結 Tower Hotel，上面有確切的地址以及詳細方位。

我們非常希望你能赴宴。如果方便的話請回覆是否參加。

謹上

保羅

To: Paul Zhang
Subject: Re: Christmas Party Invitation

Dear Paul,

Thank you for you the e-mail. Thank you, too, for the invitation to your company's Christmas party on December 23.

My wife and I are planning to attend. Please put us down for two people.

Thank you once again. Happy holidays!

Yours truly,

Ken

中文翻譯

收件人：張保羅
主旨：回覆：聖誕派對邀請

親愛的保羅：

感謝你的來信。也感謝你邀請我參加你們公司十二月二十三日的聖誕派對。

我太太和我都打算參加。請幫我們登記兩位。

再次感謝你。佳節愉快！

謹上

肯恩

V. 如何用 e-mail 邀請參加生日派對與回覆

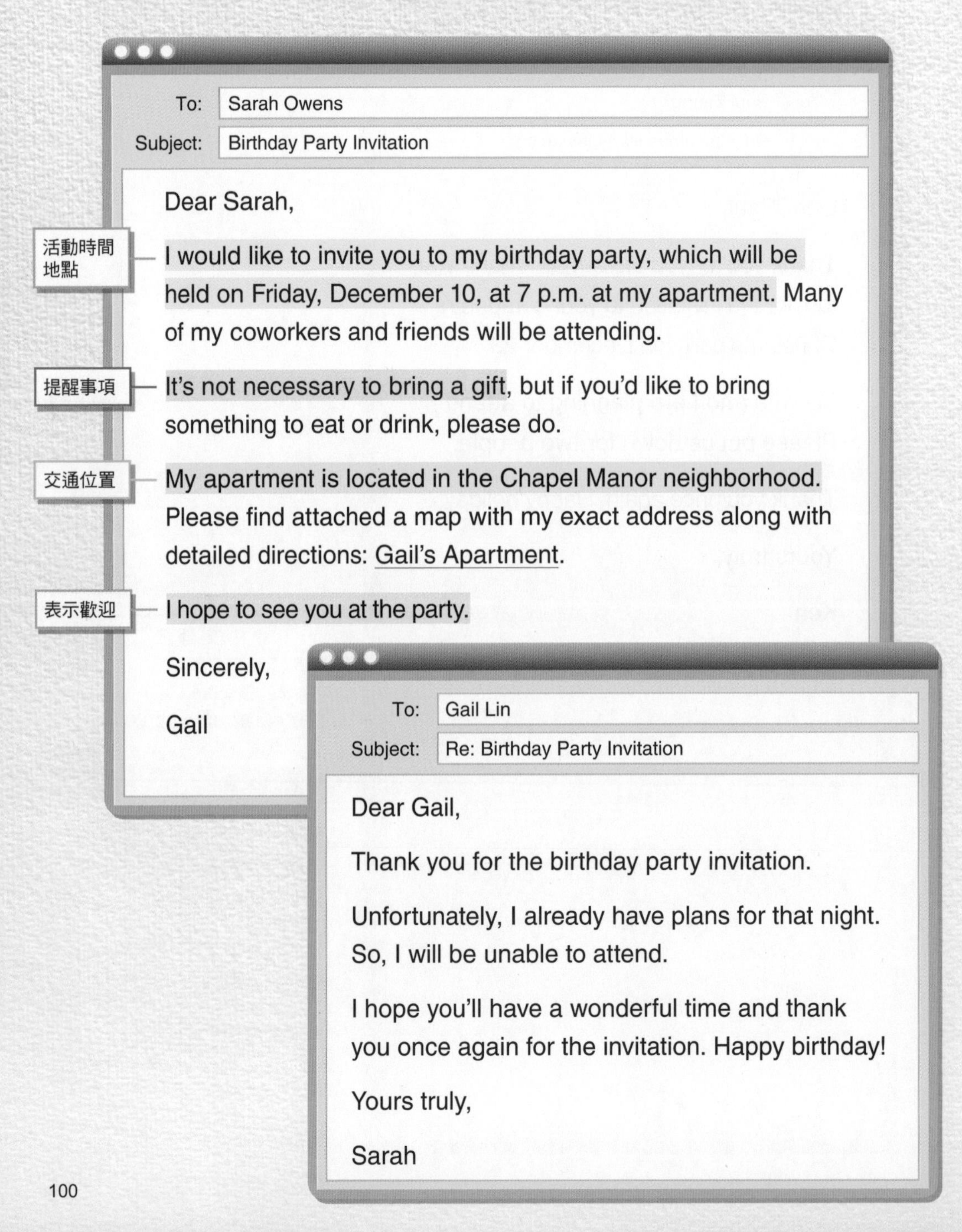

To: Sarah Owens

Subject: Birthday Party Invitation

Dear Sarah,

活動時間地點 — I would like to invite you to my birthday party, which will be held on Friday, December 10, at 7 p.m. at my apartment. Many of my coworkers and friends will be attending.

提醒事項 — It's not necessary to bring a gift, but if you'd like to bring something to eat or drink, please do.

交通位置 — My apartment is located in the Chapel Manor neighborhood. Please find attached a map with my exact address along with detailed directions: Gail's Apartment.

表示歡迎 — I hope to see you at the party.

Sincerely,

Gail

To: Gail Lin

Subject: Re: Birthday Party Invitation

Dear Gail,

Thank you for the birthday party invitation.

Unfortunately, I already have plans for that night. So, I will be unable to attend.

I hope you'll have a wonderful time and thank you once again for the invitation. Happy birthday!

Yours truly,

Sarah

中文翻譯

收件人：莎拉．歐文斯
主旨：生日派對邀請

親愛的莎拉：

我想要邀請妳來參加我的生日派對，訂於十二月十日星期五晚上七點在我的公寓舉行。我的許多同事和朋友都會參加。

不需要帶禮物，但是如果妳想帶一些東西來吃喝的話，請便。

我的公寓在教堂莊園社區裡。請見附件 Gail's Apartment 裡的地圖，裡頭有我確切的地址以及詳細的方位。

我希望能在派對上見到妳。

謹上

蓋爾

收件人：林蓋爾
主旨：回覆：生日派對邀請

親愛的蓋爾：

感謝你的生日派對邀請。

很遺憾，那天晚上我已經有計畫了。所以我沒有辦法參加。

我希望你玩得開心，並再次感謝你的邀請。生日快樂！

謹上

莎拉

Try it! 換你試試看!

1. 我想要邀請你和你太太來參加我們感恩節晚餐派對。

2. 婚禮將在二〇一二年六月十五日於 Royal Regency 飯店舉行。

3. 沒有必要帶禮物。

4. 我們希望在這個特別的日子你可以和我們一起參加。

答案請參閱第 367 頁

VI. 邀請參加活動一定要會這樣說

MP3 TRACK 52

活動時間地點

A **The party will be held on** Friday night **at** 7 p.m.

派對會在星期五晚上七點舉行。

B I look forward to being there.

我很期待出席。

A **The wedding will take place next month** at the Catholic Church.

婚禮將在下個月於天主教堂舉行。

B Great. Do you know the time and the date of the wedding?

太棒了。你知道婚禮的時間跟日期嗎？

服裝規定

A **The dress code for the party will be** formal.

須穿著正式服裝出席派對。

B Thank you. That's good to know.

謝謝。很高興知道這件事。

注意事項

A **Please feel free to bring some** food or drinks **to** the party.

想帶一些食物和飲料來參加派對的話請自便。

B That's a great idea. I'll bring my famous curry.

那是個好主意。我會帶我最有名的咖哩。

交通位置

A **Here is a map with detailed directions to the event.**

這裡是一份活動會場詳細方位的地圖。

B Thanks. This is very useful.

謝謝。這非常有用。

感謝邀請

A **Thanks for inviting me to** your party.

謝謝你邀請我參加派對。

B Well, I certainly hope to see you there.

嗯，我非常希望能在那裡看到你。

VII. 如何用電話來邀請別人

丹妮絲打電話給豪爾，邀請他來參加她的生日派對。

LISTENING 請聽 MP3 TRACK 53 ☐ | SPEAKING 請跟著 MP3 唸唸看 ☐

Denise: Hi, I'd like to speak to Hal Parsons, please.

Hal: This is Hal. Is this Denise Waitley?

Denise: It sure is. I'm calling to invite you to my birthday party.

Hal: Thanks, Denise. Where and when will it be?

活動時間地點 — Denise: The party will be at my house, this coming Friday, at 7:00 p.m.

Hal: Sounds good. I'm sure I can make it. Is there something in particular you'd like me to bring?

注意事項 — Denise: If you'd like, you could bring a bottle of wine, but it's really not necessary to bring a gift.

Hal: Could you remind me where you live?

交通位置 — Denise: I'll send you an e-mail this evening with my address and driving directions.

Hal: Great. Thanks again, Denise. I'll see you at the party.

嗨，麻煩一下我要找豪爾·帕森斯。

我是豪爾。妳是丹妮絲·維特利嗎？

是的。我打電話是要邀請你來參加我的生日派對。

謝謝妳，丹妮絲。是在什麼地方跟什麼時候呢？

派對在我家舉行，在這個星期五晚上七點。

聽起來不錯。我確定能參加。妳要我帶什麼特別的東西嗎？

如果你要的話，可以帶一瓶紅酒，不過真的不需要帶禮物。

妳可以提醒我妳家在哪裡嗎？

今晚我會寄電子郵件給你，上面會有我的地址跟行車路線。

太好了。再次感謝妳，丹妮絲。派對上見。

Try it! 換你試試看!

WRITING	請依提示寫出完整句子	☐
LISTENING	請聽 MP3 TRACK 54	☐
SPEAKING	請跟著 MP3 唸唸看	☐

1. Ⓐ Hi, James, what can I do for you?

   Ⓑ ______________________________

   ____

   ____

   (calling / invite / company / Christmas party)

2. Ⓐ When and where will the party be held?

   Ⓑ ______________________________

   ____

   ____

   (party / be / held / my apartment / Friday / 7 p.m.)

3. Ⓐ Is there dress code for the wedding?

   Ⓑ ______________________________

   ____

   (dress code / wedding / formal)

4. Ⓐ Thanks for letting me know about the party.

   Ⓑ ______________________________

   ____

   (hope / see / there)

答案請參閱第 367 頁

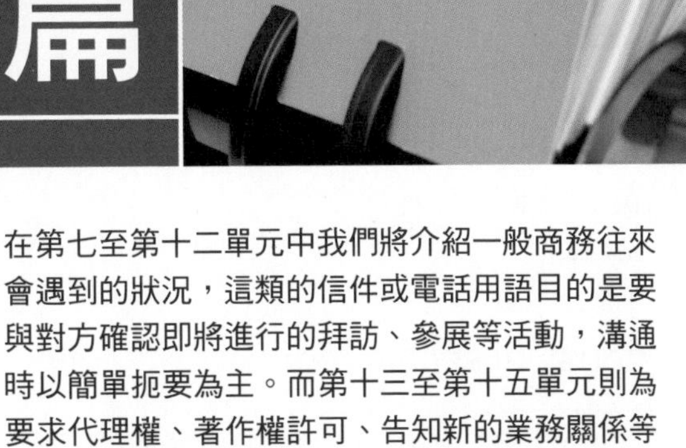

商務往來篇

在第七至第十二單元中我們將介紹一般商務往來會遇到的狀況，這類的信件或電話用語目的是要與對方確認即將進行的拜訪、參展等活動，溝通時以簡單扼要為主。而第十三至第十五單元則為要求代理權、著作權許可、告知新的業務關係等情況，需多留意專有名詞以及合約細節的說法。

UNIT
07 展開業務關係
Starting a Business Relationship

I. 展開業務關係一定要會的單字片語

MP3 TRACK 55

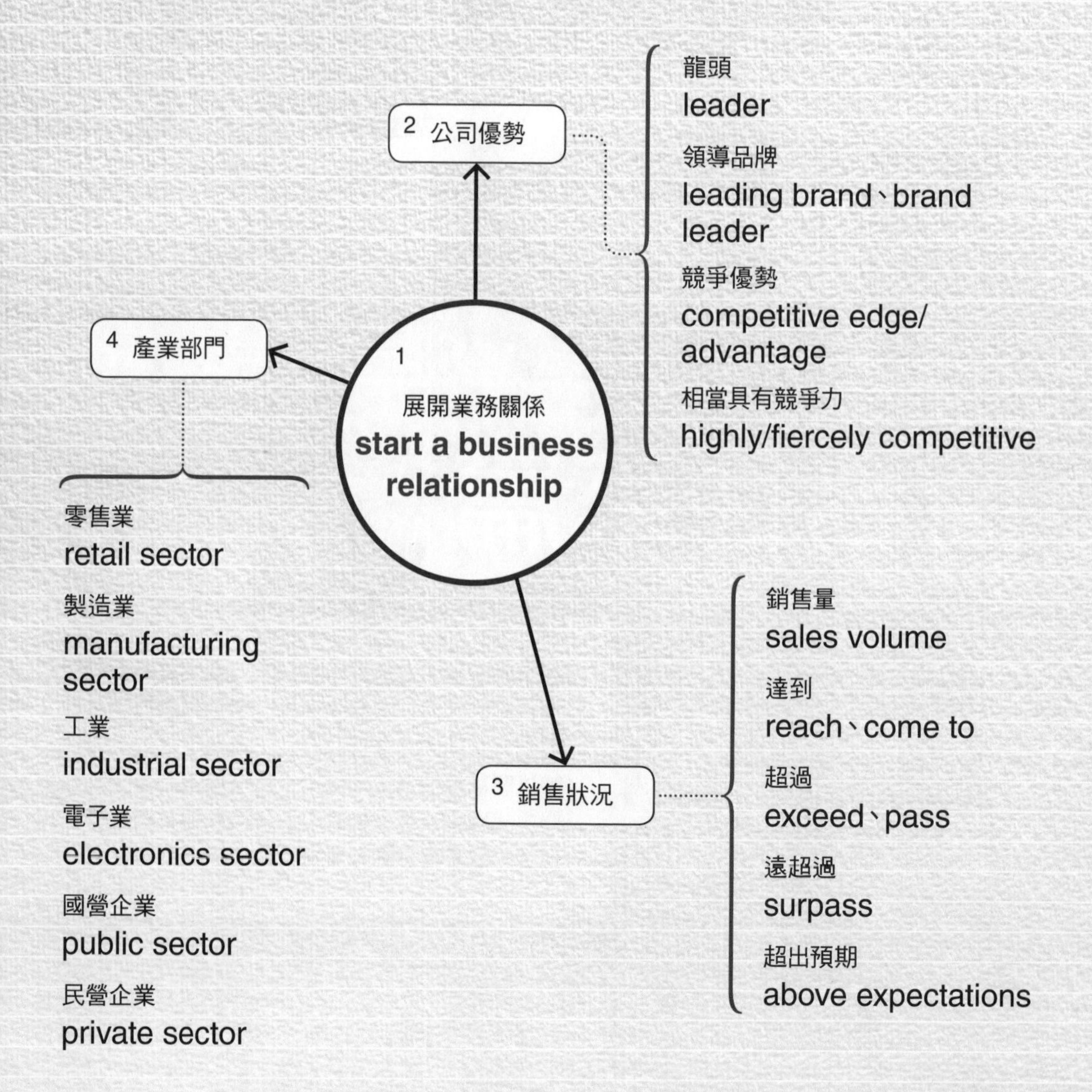

Ⅱ. 展開業務關係一定要會的句型

句型1 自我介紹（公司）

We produce sth（產品）**with** sth（技術優勢等）

例 We produce custom-built LCD screens with our state-of-the-art manufacturing facilities here in Taiwan.
我們用台灣這裡最先進的生產設備製造客製化液晶螢幕。

句型2 產品銷售狀況

Our sales volume surpassed/exceeded + 數量（+ 時間）

例 Our sales volume surpassed one million units last year, making us the largest supplier of LCD screens in the region.
我們去年銷售量遠超過一百萬台，使我們成為本地液晶螢幕最大的供應商。

句型3 建議對方應瞭解的公司情況

You should know that S. + V.

例 You should know that we have a reputation for service and reliability.
您應該要知道的是我們在服務以及信賴度方面享有盛名。

句型4 期盼有機會合作、共創事業

. . . **we welcome you to be part of our success.**

例 Our business is growing fast and we welcome you to be a part of our success.
我們的事業突飛猛進，歡迎您和我們一起共創成功的未來。

Ⅲ. 如何用 e-mail 展開業務關係

To: Greg Reynolds

Subject: Hello from Royal LCD

Dear Mr. Reynolds,

Brad Cooper, our mutual business associate over at Upim Technologies, suggested that I contact you to let you know about my company, Royal LCD. We produce custom-built LCD screens with our state-of-the-art manufacturing facilities here in Taiwan. 〔介紹公司〕

上班族加油站

state-of-the-art 用來形容當下最先進技術的設備或科技。

Royal LCD is a leader in the touch screen market. We have earned the respect of local and international companies by our attention to detail and high standards. We believe our finished product should be exactly what our clients order and free of defects. Our sales volume surpassed one million units last year, making us the largest supplier of LCD screens in the region. 〔銷售量〕

〔公司優勢〕 You should know that we have a reputation for service and reliability. When working with Royal LCD you can expect to receive "The Royal Treatment." All our clients benefit from the attention to detail their order deserves. Please feel free to browse through our Web site at www.royal-lcd.com.tw for more information.

〔希冀合作〕 Our business is growing fast and we welcome you to be a part of our success. Limitless Technologies can enjoy the advantages of high-quality LCD screens produced the Royal way. Our testing facilities ensure the finished product will perform as expected.

Thank you for taking the time to consider Royal LCD. Should you have any questions or would like a product demonstration, please do not hesitate to contact me.

Sincerely,

Paul Wang
Royal LCD

中文翻譯

收件人：葛瑞格．雷諾茲
主旨：皇家液晶顯示 器向您問候

雷諾茲先生您好：

我們在優平科技共同的商業夥伴布萊德．庫柏，建議我和您聯繫，讓您瞭解敝公司皇家液晶顯示器。我們用台灣這裡最先進的生產設備製造客製化的液晶螢幕。

皇家液晶顯示器是觸控式螢幕市場的龍頭。我們以注重細節及對高標準的要求而獲得國內外公司的推崇。我們相信我們的成品絕對符合客戶需求並且品質無虞。我們去年銷售量遠超過一百萬台，使我們成為本地液晶螢幕最大的供應商。

您應該要知道的是我們在服務及信賴度方面享有盛名。與皇家液晶螢幕顯示器合作，您可以預期能得到「皇家般的禮遇」。我們所有客戶皆因我們對商品細節用心而受益。想瞭解更多的訊息，歡迎隨時至我們網站（www.royal-lcd.com.tw）瀏覽。

我們的事業突飛猛進，歡迎您和我們一起共創成功的未來。無限科技可以享受到皇家生產製造的高品質液晶顯示器帶來的好處。我們的測試設備確保成品表現會與預期相同。

感謝您撥冗考慮皇家液晶顯示器。倘若您有任何問題或需要產品介紹的話，請儘管和我聯絡。

謹上

王保羅
皇家液晶顯示器

Ⅳ. 如何用 e-mail 回覆展開業務關係的信函

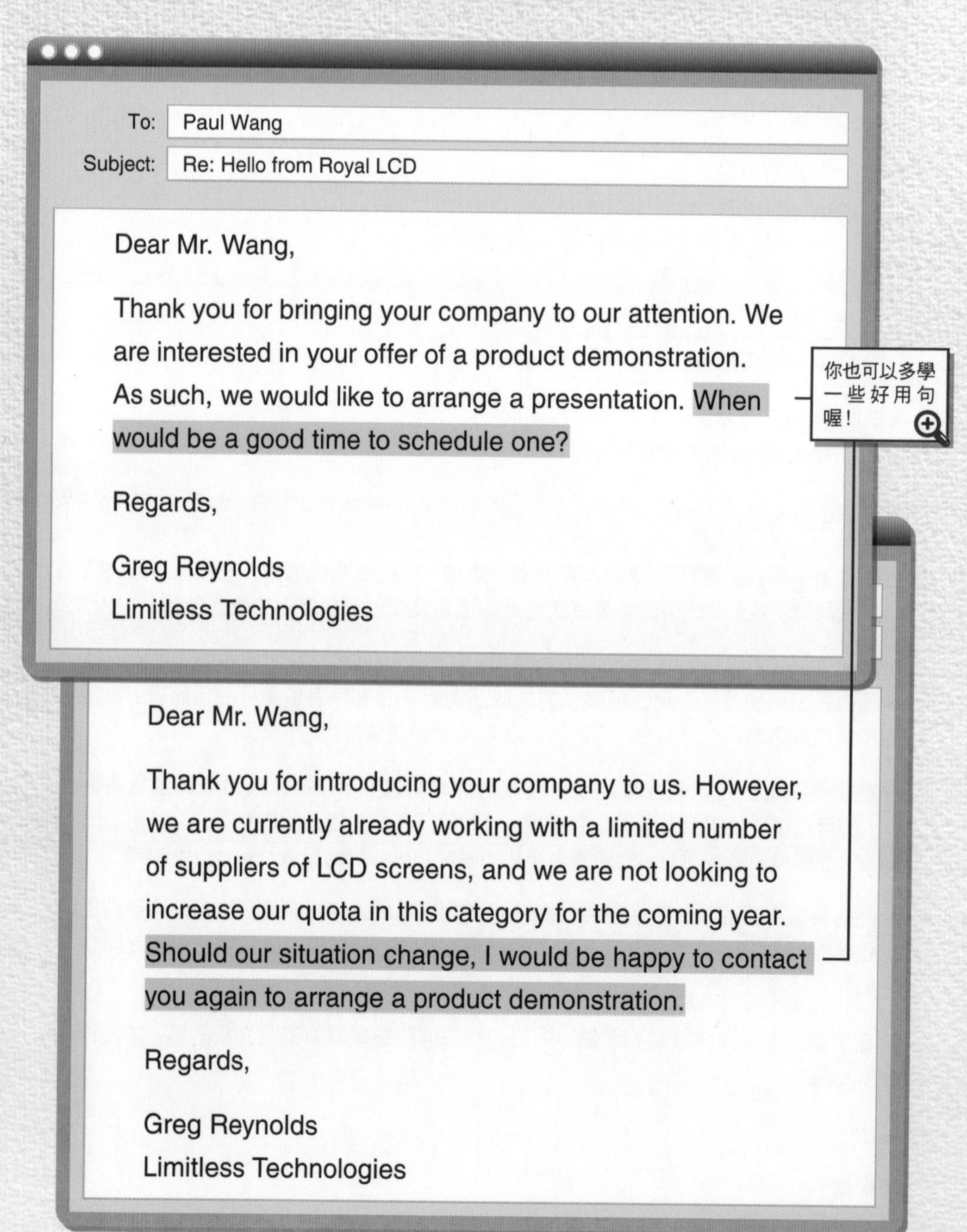

To: Paul Wang

Subject: Re: Hello from Royal LCD

Dear Mr. Wang,

Thank you for bringing your company to our attention. We are interested in your offer of a product demonstration. As such, we would like to arrange a presentation. When would be a good time to schedule one?

Regards,

Greg Reynolds
Limitless Technologies

Dear Mr. Wang,

Thank you for introducing your company to us. However, we are currently already working with a limited number of suppliers of LCD screens, and we are not looking to increase our quota in this category for the coming year. Should our situation change, I would be happy to contact you again to arrange a product demonstration.

Regards,

Greg Reynolds
Limitless Technologies

中文翻譯

收件人：王保羅
主旨：回覆：皇家液晶顯示器向您問候

王先生你好：

謝謝你告訴我們貴公司的一些事情。我們對於你所提議的產品展示感興趣。如上所言，我們希望可以安排一場簡報。安排在什麼時候會比較方便呢？

謹上

葛瑞格‧雷諾茲
無限科技

王先生你好：

感謝你向我們介紹貴公司。然而，我們目前已經和固定幾家的液晶顯示器供應商合作，並且我們在未來一年並沒有預期要增加這個類別的配額。倘若我們的情況有改變，我將很樂意再與你聯繫來安排產品展示會。

謹上

葛瑞格‧雷諾茲
無限科技

Try it! 換你試試看!

1. 我們提供香港、新加坡、台灣和菲律賓多種高品質的文具用品。

(多種的 a wide range of　文具用品 stationery)

2. 我們上一季的銷售額超過八百萬。

(銷售額 sales volume)

3. 你應該知道我們公司去年榮獲亞洲最佳企業獎。

4. 歡迎你成為我們快速發展事業中的一員。

(快速發展的 fast-growing)

答案請參閱第 368 頁

V. 展開業務關係一定要會這樣說

MP3 TRACK 56

介紹公司

A	What does your company do?	你們公司是做什麼的？
B	**We supply** high-performance transistors **to** the international marketplace.	我們供應高效能的電晶體給國際市場。

介紹公司主產品

A	What does your company produce?	你公司是生產什麼的？
B	**My company produces** solar modules for commercial application.	我公司生產作為商業用途的太陽能模組。

產品銷售狀況

A	What is your sales volume?	你們銷售額是多少？
B	**Last year we sold more than** 100,000 units.	去年我們賣出超過十萬組。

回答也可以說 Our sales volume reached 100,000 units last year.

產能狀況

A **How many units did you turn out** last year?

你們去年的產能表現如何？

B We shipped out over 50,000 units last year alone.

光是去年我們的出貨量就超過五萬套。

介紹公司優勢

A What else can you tell me about your company?

關於你們公司還有什麼可以告訴我的嗎？

B **You should also know that** we provide a money-back guarantee.

您應該也要知道我們提供退款保證。

歡迎試用

A **We would like to welcome you to sample our product.**

歡迎您試試我們的產品。

B We will definitely consider your offer.

我們一定會考慮你的提議。

VI. 如何展開業務關係

環球實業的李漢克打電話給一位潛在客戶。

LISTENING 請聽 MP3 TRACK 57 □ SPEAKING 請跟著 MP3 唸唸看 □

Hank: Hello, is this Ms. Owen? This is Hank Lee from Global Industries.

Tanya: How can I help you, Mr. Lee?

Hank: I'm calling to introduce my company and our services. I think you'll be interested to hear what we have to offer.

Tanya: OK. What does your company do?

介紹公司

Hank: We produce hand-held power tools for the retail sector.

Tanya: I see. What else can you tell me?

公司優勢

歡迎合作

Hank: You should know that all our products come with a money-back guarantee. Not only that, but we sold over 200,000 power drills just in the past two months.

Tanya: You don't say! You've sparked my interest, Mr. Lee. Why don't you leave your contact information with my secretary and we can schedule a product demonstration?

Hank: Excellent. I look forward to meeting you in person.

Tanya: As do I. I'll transfer you over now.

Hank: Thank you. Have a great day!

Tanya: You too. So long.

您好，請問是歐文小姐嗎？這裡是環球實業的李漢克。

李先生，有什麼能效勞的嗎？

我來電是想要介紹敝公司以及我們的服務項目。我想您會有興趣聽聽我們的提議。

好吧。你們公司是做什麼的？

我們為零售商生產手持電動工具。

我瞭解了。還有其他可以告訴我的嗎？

您應該知道我們所有產品都有退款保證。不僅如此，光是過去兩個月我們就賣出超過二十萬支的電鑽。

真的假的？李先生你引起我的興趣了。你不妨把聯絡資料留給我秘書，然後我們可以安排一場產品說明會。

太好了。我期待能見到您本人。

我也是，我現在就幫你轉過去。

謝謝。祝您有個美好的一天。

你也是。再見。

Try it! 換你試試看!

WRITING	請依提示寫出完整句子	☐
LISTENING	請聽 MP3 TRACK 58	☐
SPEAKING	請跟著 MP3 唸唸看	☐

1. Ⓐ What exactly does your company do?

   Ⓑ ______________________________

   ____

   (produce / solar panels / industrial sector)

2. Ⓐ How many units did you sell last year?

   Ⓑ ______________________________

   ____

   (sales volume / pass / 400,000 units)

3. Ⓐ What are some of your company's achievements?

   Ⓑ ______________________________

   ____

   (won / Marshall Award / three years in a row)

4. Ⓐ When can we have a look at your product samples?

   Ⓑ ______________________________

   ____

   (invite / attend / product demonstration)

答案請參閱第 368 頁

UNIT 08 確認業務拜訪

Confirming a Business Meeting

I. 確認業務拜訪一定要會的單字片語

MP3 TRACK 59

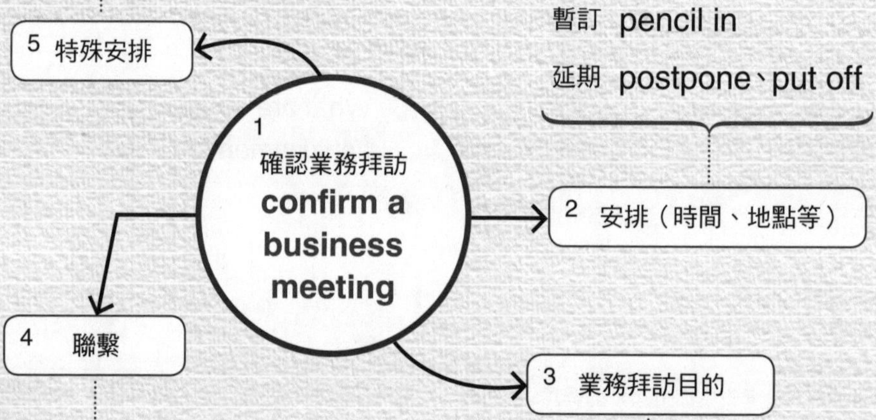

1 確認業務拜訪 confirm a business meeting

2 安排（時間、地點等）

安排時間 schedule、time

重新計畫、安排 rearrange、reschedule

敲定 fix

暫訂 pencil in

延期 postpone、put off

3 業務拜訪目的

介紹最新產品 introduce the latest products

產品展示 give a product demonstration

提供優惠方案 offer a special deal

4 聯繫

再次確認 reconfirm

（跟相關人員）進行聯繫 touch base with sb

聯繫上某人 check in with sb、get in touch with sb、get a hold of sb

5 特殊安排

液晶投影機 LCD projector

投影機 overhead projector

螢幕 screen

簡報 PowerPoint presentation

Ⅱ. 確認業務拜訪一定要會的句型

句型1 提醒拜訪及此行目的

I just wanted to remind you of my upcoming visit to your company + 目的

例 I just wanted to remind you of my upcoming visit to your company to introduce our company's latest products.
我只是想要提醒你，我即將去你們公司拜訪並介紹我們公司的最新產品。

句型2 確認時間地點

To confirm, I am scheduled to meet with you + 時間、地點

例 To confirm, I am scheduled to meet with you at 10 a.m. next Friday, December 30, at your office.
確認一下，我預定在十二月三十日下星期五早上十點在你們辦公室和你見面。

句型3 設備需求

I would also appreciate it if you could + V.

例 I would also appreciate it if you could arrange to have a screen and LCD projector in the conference room to facilitate my presentation.
如蒙預先準備螢幕和液晶投影機以方便進行簡報的話，則不勝感激。

句型4 下次聯絡時間

I will contact you + 時間 + **to reconfirm details.**

例 I will contact you in the next few days to reconfirm details.
過幾天我再和你聯繫，再次確認細節。

Ⅲ. 如何用 e-mail 確認業務拜訪日期

To: Ken Burns

Subject: Reminder: December 30 meeting

Dear Ken,

拜訪提醒

I just wanted to remind you of my upcoming visit to your company to introduce our company's latest products. To confirm, I am scheduled to meet with you at 10 a.m. next Friday, December 30, at your office.

確認時間地點

Please note that I have already mailed you our company's most recent catalog, so you can preview any of these items before we meet.

設備需求

I would also appreciate it if you could arrange to have a screen and LCD projector in the conference room to facilitate my presentation. If this is not possible, please let me know, and I will bring my own portable equipment.

下次聯絡時間

I will contact you in the next few days to reconfirm details.

Thank you for your time. I look forward to meeting with you soon.

Kind regards,

Harry Wu, Sales Representative
Wisegrowth, Inc.
333-134-5676 ext. 146 (office)
445-786-3476 (cell)

中文翻譯

收件人：肯恩．波恩斯

主旨：提醒十二月三十日的會面

親愛的肯恩：

我只是想要提醒你，我即將到貴公司拜訪並介紹敝公司的最新產品。確認一下，我預定會在十二月三十日下個星期五早上十點在你們辦公室和你見面。

請留意我已將敝公司最新的型錄寄給你，所以在我們見面前你可以事先看一下這些商品。

如蒙預先準備螢幕和液晶投影機以方便進行簡報的話，則不勝感激。如果不可行，請讓我知道，我會把我的攜帶式設備帶去。

過幾天我再和你聯繫，再次確認細節。

感謝你撥出寶貴的時間。我期待盡快和你見面。

謹上

業務代表　吳哈利
明智發展公司
333-134-5676 分機 146（公司電話）
445-786-3476（手機）

文化補給站

每個國家做生意的方式都不太一樣，當你要去其他國家進行業務拜訪時，應該要先瞭解對方做生意的方式及風格，例如對時間的觀念在不同文化之間就有很大的差異，一般來說有單向的（monochromic）及多向的（polychromic）兩種。單向的時間觀是線性的，做事時喜歡一次專注在一件事情上，如美國、德國、日本、或是以鐘錶製造聞名世界的瑞士。而多向的時間觀則表示他們習慣在同一時間內從事多項事情，同時牽涉不同的人，這種態度比較常見於中東國家、法國、義大利、希臘等拉丁國家，以及某些東方和非洲文化的國家。

單向時間觀的人在進行會議時傾向於：

- 比較喜歡迅速開始或結束，不喜歡拖泥帶水
- 依照時間表休息
- 一次處理一件議程
- 依序發言
- 視遲到為不尊重的表現

多向時間觀的人進行會議時傾向於：

- 對於會議的開始及結束都很彈性
- 在會議進行差不多時就會稍作休息
- 對於同時出現許多的訊息習以為常
- 經常彼此交談或是打斷對方的談話
- 不認為遲到是冒犯的行為

所以，與不同文化的人做生意時，要想想他們是從哪裡來的，才不會覺得他們的行為不尊重人或是不通人情。

Ⅳ. 如何用 e-mail 回覆業務拜訪確認函

To: Harry Wu

Subject: Re: Reminder: December 30 meeting

Dear Harry,

Thanks for the reminder[1] about your upcoming visit to our company. I have received your catalog, and I look forward to learning more about your company's newest products. I will make sure that the items you've requested[2] are set up in the conference room prior to your arrival.

I will see you next Friday at 10 a.m.

Sincerely,

Ken

上班族加油站

prior to 為介系詞片語，表示「在……之前」的意思，為正式的用法。

Vocabulary & Phrases

1. reminder [rɪ`maɪndɚ] *n.*
作為提醒的人事物

2. request [rɪ`kwɛst] *v.*
請求；拜託

中文翻譯

收件人：吳哈利
主旨：回覆：提醒十二月三十日的會面

親愛的哈利：

感謝你提醒即將來我們公司拜訪的事。我有收到你們的型錄，並且期待能更瞭解你們公司的最新產品。我會確認你所要求的這些配備在你到達之前會在會議室裡設置好。

下個星期五早上十點見。

謹上

肯恩

Try it! 換你試試看!

1. 只是要提醒你，我即將去你們公司拜訪並介紹我們公司的最新產品。

2. 我想要確認我們會議訂於十二月三十日星期五早上十點在你的辦公室進行。

3. 有沒有可能要求在簡報時使用投影機？

4. 我下星期會再跟你聯繫。

答案請參閱第 368 頁

V. 確認業務拜訪一定要會這樣說

MP3 TRACK 60

提醒拜訪時間

A Are we supposed to meet next week?

我們下星期是不是應該要見面？

B Yes. **We are scheduled to** get together next week to review the changes to the contract.

是的。我們預定下星期要一起再討論合約修改的地方。

確認時間跟地點

A **I just wanted to confirm that** we will be meeting at your company tomorrow morning at nine a.m.

我只是要確認明天早上九點我們會在你公司見面。

B That's correct. Just give your name to the receptionist when you arrive.

沒有錯。你到的時候只要告訴櫃臺人員你的名字就可以了。

A Is our meeting in two weeks still on?

我們兩個星期後的會議還會進行嗎？

B Yes. **We are scheduled to meet** at my office at two o'clock, two weeks from now.

會。我們預定兩個星期後兩點的時候在我辦公室見面。

設備需求

A	**I'd appreciate your arranging some presentation equipment for our meeting** next week.	如蒙預先準備我們下星期會議使用的簡報設備，則不勝感激。
B	No problem. What exactly do you need?	沒問題。確切來說你需要什麼？
A	**Would it be possible to have access to a** projection screen at the presentation venue?	簡報會場有沒有可能可以使用投影螢幕？
B	I'm afraid that's not possible. We only have LCD monitors.	恐怕不太可能。我們只有液晶螢幕。

商務往來篇

下次聯絡時間

A	**I'll check in with you again** next week **to reconfirm the meeting.**	下星期我會再跟你聯繫，再次確認會面事宜。
B	Great. Talk to you again soon.	太好了。我們再聯絡。

Talk to you again soon. 為掛電話時的用語，表示「再聯絡」的意思，也可以說 Talk to you later。

VI. 如何用電話確認業務拜訪

基因科技公司的林蘇打電話給約翰·史蒂爾確認業務拜訪的時間。

LISTENING 請聽 MP3 TRACK 61 ☐ | SPEAKING 請跟著 MP3 唸唸看 ☐

Sue: Hello. I'm calling for John Steele.

John: John speaking. How can I help you?

拜訪提醒

Sue: This is Sue Lin over at GeneTech. I'm calling about my visit to your company next week to introduce our latest products. I just wanted to make sure that it's still on.

John: Oh, that's right. When exactly are we meeting again?

確認時間地點

Sue: We are scheduled to meet next Tuesday at 10:30 a.m. at your office.

John: Yes, I have it written down in my planner here. I'm looking forward to it.

Sue: Me too. And I'm certain that you will find our new products beneficial to your company.

John: Thanks for the reminder, Sue. I'll see you next Tuesday.

設備安排

Sue: Great. By the way, I'd like to know if it would be possible to request some presentation equipment in the room where we'll be meeting.

John: No problem. Just e-mail me a list of what you need, and I'll make sure it's there.

下次聯絡時間

Sue: That's great, John. Thank you. I'll call you again next week to firm things up.

你好。我要找約翰·史蒂爾。

我是約翰。我可以幫妳什麼？

我是基因科技公司的林蘇。我來電是關於下星期我要去貴公司介紹我們最新的產品。我只是要確定這事還在進行中。

喔，沒錯。我們下次見面的確切時間是什麼時候？

我們預定下星期二早上十點半在你辦公室見面。

好的，我記在行事計畫表上了。我很期待會面。

我也是。而且我確信你會發現我們產品對你們公司有幫助。

蘇，感謝妳的提醒，我們下星期二見。

太好了。對了，我想知道是否有可能在我們開會的房間要求一些簡報用的設備。

沒問題。妳只要把需要清單寄給我，我會確認設備都安裝好。

那太棒了，約翰。謝謝。我下星期會再打電話把事情確定下來。

Try it! 換你試試看!

WRITING 請依提示寫出完整句子	☐
LISTENING 請聽 MP3 TRACK 62	☐
SPEAKING 請跟著 MP3 唸唸看	☐

1. Ⓐ Hi Angela. What can I do for you?

   Ⓑ ______________________________

   ____

   (be calling / remind / upcoming visit / your company / introduce / latest products)

   Ⓐ Thanks for reminding me. I'm looking forward to it.

2. Ⓐ I think we are scheduled to meet sometime soon. Can you confirm this?

   Ⓑ ______________________________

   ____

   (confirm / I am scheduled / meet / next Friday / 2 p.m. / office)

3. Ⓐ Do you have any other questions?

   Ⓑ ______________________________

   ____

   (yes / would / know / projection screen / conference room)

   Ⓐ Of course. Our conference room is state-of-the-art.

4. Ⓐ I think we're all set to meet next Monday.

   Ⓑ I'm really looking forward to it. __________

   ____

   ____

   (contact / again / Friday / reconfirm / everything)

答案請參閱第 368 頁

UNIT

09 邀請蒞臨貿易展

Invitation to a Trade Show

I. 邀請蒞臨貿易展一定要會的單字片語

MP3 TRACK 63

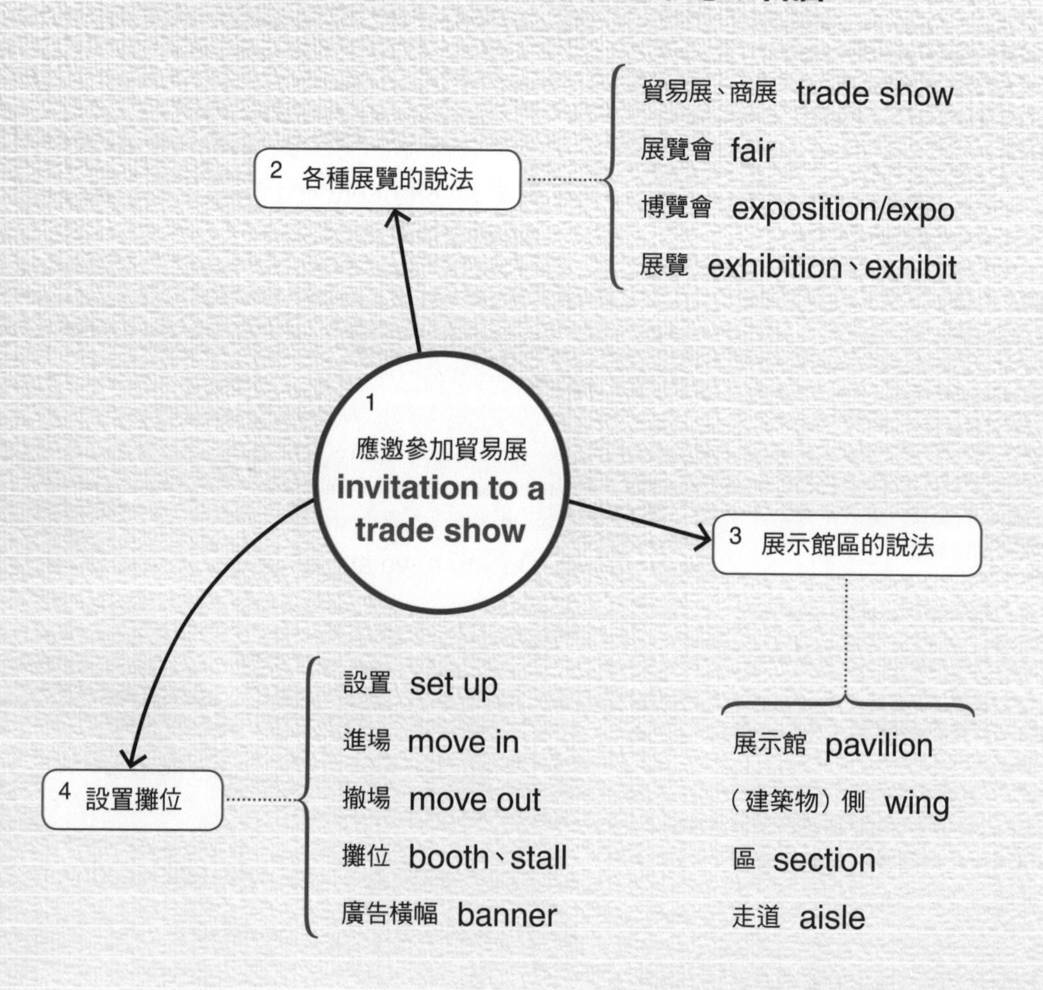

Ⅱ. 邀請蒞臨貿易展一定要會的句型

句型 1　提出邀請

S. **is delighted to invite you to take a look at our booth** + 時間 + 貿易展

例 Ovid Book Publishers is delighted to invite you to take a look at our booth at next month's Asian International Book Fair.
奧維德圖書出版公司很高興邀請你下個月到我們亞洲國際書展的攤位參觀。

句型 2　展場位置

Our booth number is + 號碼**, and we will be set up in** + 位置

例 Our booth number is 254, and we will be set up in the West Wing of Pavilion 2.
我們的攤位編號是二五四，將設在展示館二館的西側。

句型 3　參觀提醒

Don't forget + to V. **for a chance** + to V.

例 Don't forget to print out the attached coupon and bring it to the show for a chance to win a free iPad.
別忘了將隨附的優待券列印出來並帶到展場，就有機會免費獲得 iPad。

Ⅲ. 如何用 e-mail 寫蒞臨貿易展的邀請函

To: Joe Dickinson

Subject: Visit us at the Asian International Book Fair (AIBF): Jan. 16–18, 2012

Dear Joe,

展出邀請 — Ovid Book Publishers is delighted to invite you to take a look at our booth at next month's Asian International Book Fair.

I will be managing the booth personally the entire day on all three days, so please drop by any time, and I would be happy to introduce our latest print products to you. I'm certain that your customers will be very interested in these new products.

What's more, you'll have the chance to get an exclusive look at our new digital educational series, which is due out on the market next February. So come join us and be one of the first to test-drive our revolutionary multimedia products.

展場位置 — Our booth number is 254, and we will be set up in the West Wing of Pavilion 2. Just look for our company's signs and banners to find us.

參觀提醒 — Lastly, don't forget to print out the attached coupon and bring it to the show for a chance to win a free iPad.

If you have any questions, please don't hesitate to contact me.

We look forward to seeing you at the AIBF.

Sincerely,

Melody Lin, Marketing Executive
Ovid Book Publishers

中文翻譯

收件人：喬．迪金森
主旨：謹於二〇一二年一月十六日至
　　　十八日蒞臨亞洲國際書展

親愛的喬：

奧維德圖書出版公司很高興邀請您下個月到我們亞洲國際書展的攤位參觀。

這三天我本人都會整天在攤位上負責處理展場事宜，所以隨時歡迎順道過來參觀，我會很樂於向您介紹我們最新的出版品。我確信您的顧客對我們這些新產品會很感興趣。

除此之外，您將有機會獨家一窺我們預計在明年二月上市的新數位教學系列。所以來加入我們，成為我們革命性多媒體產品的首批體驗者吧。

我們的攤位號碼是二五四，將設在展示館二館的西側。只要尋找我們公司的標誌和廣告橫幅就能找到我們。

最後，別忘了將隨附的優待卷列印出來並帶到展場，就有機會免費獲得 iPad。

如果您有任何問題的話，請別客氣儘管打電話給我。

我們期盼在亞洲國際書展見到您。

謹上

行銷經理　林美樂蒂
奧維德圖書出版公司

文化補給站

發送貿易展邀請函的小訣竅

電子郵件對於推廣活動或促銷商品的效果其實不容小覷，不過要先掌握下列這幾個要點：

- **寄送對象**
  只邀請對你們展示商品有興趣的客戶，包括重要的潛在客戶，以及方圓一百英里內的潛在客戶。
- **優惠券** (gift coupon)
  隨函附贈優惠券，可到現場兌換精美小禮物或參加抽獎。
- **以個人的名義發函**
  用個人名義署名的話，收信人感覺會較親切，效果也會比較大。
- **首度獨家公開** (exclusivity)
  強調到現場可先睹為快，且強調獨家性通常帶有誘因，如範文中提到的 have the chance to get an exclusive look at . . .。
- **利用多重部分的發信系統**（multi-part mailing system）
  通常連續發送多封信件的效果會比只發送一封的效果來得大。你可以將訊息分成下列三個部分：
  1. 第一封信只預告活動訊息，通常可以引起收件人的興趣和好奇。
  2. 第二封則包含完整的套裝邀請，包括活動訊息、優惠券、手冊、導覽等。
  3. 第三封可以是後續的提醒函，若對方已回覆參加，則可以確認活動的日期、時間、地點等。

Ⅳ. 如何用 e-mail 回覆是否蒞臨貿易展

To: Melody Lin

Subject: Re: Visit us at the Asian International Book Fair (AIBF) : Jan. 16–18, 2012

Dear Melody,

Thank you for the invitation. I have noted your booth number and location at the AIBF, and I will be sure to drop by.

I am also very interested in trying out your new digital products, and seeing how they match up with the products offered in the U.S. If possible, please e-mail me an information and pricing catalog so that I can get a preview before coming to the show.

Well, it'll be a pleasure to see you at the show. I'll be sure to e-mail you if I have any questions.

Sincerely,

Joe

上班族加油站

drop by 表示「順道拜訪」，也可以用 drop in、come over、stop by、stop in 等動詞哦。

中文翻譯

收件人：林美樂蒂
主旨：回覆：謹於二〇一二年一月十六日至十八日蒞臨亞洲國際書展

親愛的美樂蒂：

感謝妳的邀請。我已經記下你們在亞洲國際書展的攤位號碼以及位置，我一定會過去參觀。

我也很有興趣試用一下你們新的數位產品，看看它們和美國所推出的產品搭配的效果如何。如果可以的話，請把資料以及價目表用電子郵件寄給我，這樣我可以在去展場之前先預看一下。

嗯，很高興即將可以在展場上見到妳。如果有任何問題的話我一定會用電子郵件跟妳聯絡。

謹上

喬

Try it! 換你試試看!

1. 我寫信是要邀請你下個月在即將來臨的台北貿易展上到我們攤位參觀。

2. 我們會在這次貿易展上展出我們最新系列的產品。

3. 我們的攤位號碼是一二五號，位於貿易三館走道 B 上。

4. 別忘了攜帶附件的優惠券，就有機會贏得大獎。

答案請參閱第 368 頁

V. 邀請蒞臨貿易展一定要會這樣説

MP3 TRACK 64

提出邀請

A I'd like to learn more about your company and its products.

我想要知道更多關於你們公司以及產品的事。

B Then, **please come and visit our booth at the trade show** this week.

那麼，請這個星期到我們貿易展的攤位來參觀。

展場活動

A Will you be giving a demo of your products at the trade show?

你會在貿易展上介紹你的產品嗎？

B Yes. **Our representatives will be demonstrating** how to use our products.

會的。我們的業務人員會展示如何使用我們的產品。

A **You'll have a chance to get a sneak preview at** the latest version of our products.

你將有機會先目睹我們產品的最新版本。

B Great to know. I'll definitely be there for that.

很高興知道這件事。我一定會為此去的。

展場位置

A In which part of the convention center will your booth be?

你們攤位在會議中心哪一區呢？

B **We are in** section A of Pavilion 3, with the rest of the multimedia companies.

我們在展示三館的A區，和其他多媒體公司在一起。

A How can we find your booth?

我們要怎麼找到你的攤位？

B **Our booth number is** 56B, and it's near the entrance of the hall.

我們的攤位號碼是56B，靠近展覽館入口處。

提醒

A **Don't forget to** present your coupon number at the booth to enter our sweepstake.

別忘了在攤位上出示優惠券的號碼，以參加我們的抽獎。

B I won't forget. And I do hope that it's a lucky number!

我不會忘記的。我希望那是個幸運號碼！

sweepstake 指一種抽獎活動，如果抽到你的名字就有機會贏得大獎項，也可用來指彩金獨得的一種比賽。

VI. 如何用電話邀請蒞臨貿易展

高唐娜打電話邀請潛在客戶來參觀她們在貿易展的攤位。

LISTENING 請聽 MP3 TRACK 65 ☐ SPEAKING 請跟著 MP3 唸唸看 ☐

Donna: Hi. Is Greg Thompson in?

Greg: Yes, this is Greg. Is this Donna Kao?

展出邀請

Donna: It sure is. I'm just calling to remind you to stop by and visit us at our booth at the upcoming Taipei Trade Show this spring.

Greg: I'm already planning to do so. Any surprises this year?

展出項目

Donna: Well, we will be introducing our latest line of energy-efficient devices at this year's event.

Greg: I've been wanting to get a closer look at those.

展場位置

Donna: You can do so if you stop by booth 88 in section C of the trade complex. That's where you'll find us.

Greg: I'm writing down your booth number now.

Donna: I'll e-mail this information to you as well. And if you have any questions, just contact me at anytime.

Greg: Sounds great. I'm looking forward to seeing you soon.

嗨。葛瑞格・湯普森在嗎？

是的，我就是葛瑞格。妳是高唐娜嗎？

沒錯。我只是打電話提醒你今年春天即將來臨的台北貿易展你要順道過來我們攤位參觀。

我已經打算要去了。今年有什麼令人驚喜的嗎？

嗯，在今年的貿易展上我們會介紹最新系列的節能裝置。

我一直想要仔細看看這些裝置。

如果你到貿易大樓 C 區八十八號攤位來參觀的話就可以看到。你可以在那裡找到我們。

我現在把妳的攤位號碼寫下來。

我也會將這個資訊寄電子郵件給你。如果你有任何問題的話，隨時打電話給我。

聽起來很棒。期待很快就可以見到妳。

Try it! 換你試試看!

WRITING	請依提示寫出完整句子	☐
LISTENING	請聽 MP3 TRACK 66	☐
SPEAKING	請跟著 MP3 唸唸看	☐

1. **A** Hi Janet. How's business going? Anything exciting happening?

   B There sure is. ______________________

   ____

   (invite / visit / stand / upcoming / International Machinery Tool Show / next month)

   A Thanks for the invitation. I'll be sure to stop by.

2. **A** What can we expect from your company at this trade show?

   B ______________________

   ____

   (launch / exciting new products / offer / hands-on demonstrations)

3. **A** Where exactly will you be located?

   B ______________________

   ____

   (find / display / booth 25 / located / Trade Hall 1)

4. **A** I'll see you at the exhibit hall.

   B ______________________

   ____

   (if / questions / feel free / contact / me / anyone / marketing department)

答案請參閱第 369 頁

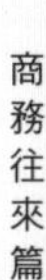

UNIT 10

安排會面

Scheduling a Meeting

I. 安排會面一定要會的單字片語

MP3 TRACK 67

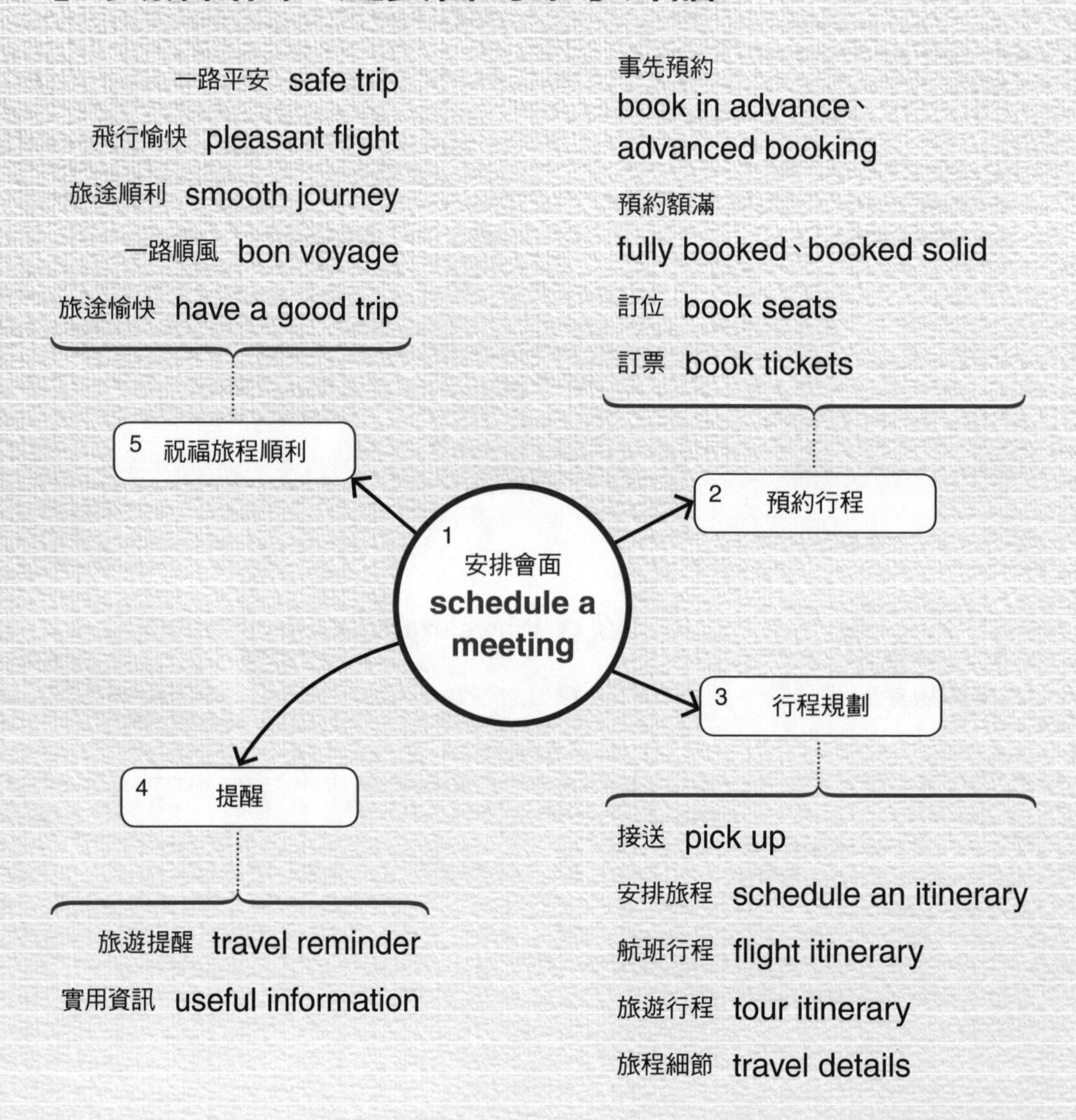

Ⅱ. 安排會面一定要會的句型

句型1 訂房

I have booked a room for you + at 地點 + for 一段時間

例 I have booked a room for you at the Regal Inn for the length of your stay.
你停留這段時間我已經幫你在富豪酒店訂了房間。

句型2 說明行程安排

I have come up with sth **for your visit.**

例 I have also come up with an itinerary for your visit.
我也為你的造訪想了一個行程計畫。

句型3 溫馨提醒

Please note that S. + V.（氣候狀況、文化習慣）

例 Please note that winter weather in Taipei can be cool and rainy.
請留意台北冬天的天氣可能又濕又冷。

句型4 旅途愉快

S. wish you + 祝福

例 We wish you a safe journey.
祝你旅途平安。

Ⅲ. 如何用 e-mail 安排會面

To: Ken Fields

Subject: Upcoming Meeting in Taiwan

Dear Ken,

I am writing to confirm the details of your upcoming visit to Taiwan from January 4 to January 7. I have booked a room for you at the Regal Inn for the length of your stay. (訂房)

As per your request, I have also come up with an itinerary for your visit. (說明行程安排) I will be at the airport to pick you up when you arrive on the evening of January 4. For our scheduled meeting on January 5, I have arranged for a car to pick you up at the hotel and bring you to our office. Since it will be your first time in Taipei, I have also organized a day of sightseeing[1] for you on January 6. For further details, please refer to the attached[2] itinerary.

Please note that winter weather in Taipei can be cool and rainy. (溫馨提醒) Also, you may want to bring some casual clothes for the city tour that has been planned.

If you need any clarifications,[3] please don't hesitate to drop me a line.[4]

We look forward to seeing you in Taipei and wish you a safe journey. (旅途愉快)

Regards,

Wes

Vocabulary & Phrases

1. sightseeing [ˋsaɪt͵siɪŋ] *n.* 觀光
2. attached [əˋtætʃt] *adj.* 附加的
3. clarification [͵klɛrəfəˋkeʃən] *n.* 說明；闡明
4. drop sb a line *v. phr.* 寫封短信

中文翻譯

收件人：肯恩・菲爾茲
主旨：即將在台灣見面

親愛的肯恩：

我寫信是要確認你將於一月四日至七日來台拜訪的細節。你停留這段時間我已經幫你在富豪酒店訂了房間。

根據你的要求，我也為你的造訪想了一個行程計畫。你一月四日晚上抵達時，我會到機場接你。關於我們預定一月五日的會面，我已經安排一部車去飯店接你到我們公司。由於這是你第一次來台北，我也替你規劃一月六日那天的一日觀光行程。關於其他更多的細節，請參閱附加檔案的行程表。

請留意台北冬天的天氣可能又濕又冷。此外，你可能會想要帶些休閒服，可以在所規劃的市區觀光行程中穿。

如果你需要任何說明，請不要客氣，儘管寫封短信給我。

我們期待在台北見到你，並祝你旅途平安。

謹上

維斯

文化補給站

送禮文化大不同

在中華文化裡，我們或許習慣在商務場合中收贈禮物，但是在其他文化中，送禮的行為並不常見。除此之外，有些公司或政府機構會明文禁止員工收禮。為了避免造成不便，和其他國家的公司有業務往來時，最好能先了解對方的習慣。

以北美來說，他們並沒有送禮的習慣，所以並不會預期在初次見面或工作場合中有人會攜帶伴手禮前來。其他沒有送禮文化的國家還包括：澳洲、丹麥、以及非洲國家。而在一些拉丁美洲國家，初次見面時不需要送禮，但是之後的會面通常就需要帶伴手禮前往。

最後再提醒你一些送禮上的禁忌：例如牛在印度是神聖的，而酒精在穆斯林文化中是被禁止的，所以如果你送牛皮製品給印度夥伴，或是帶一瓶美酒給穆斯林的客戶，都是不尊重對方文化的行為，會令對方感到被冒犯。

Ⅳ. 如何用 e-mail 回覆會面安排事宜

To: Wes Ko

Subject: Re: Upcoming Meeting in Taiwan

Dear Wes,

Thank you for confirming the dates and sending me the details of my upcoming trip to Taipei. I am very much looking forward to this visit.

If any questions come up, I'll be sure to let you know.

Thanks again for the information. I will see you in Taipei on January 4.

Sincerely,

Joe

上班族加油站

if any questions come up 與 if you have any questions 意思相近，come up 表示「（問題、困難等）出現、發生」，主詞也可以是某個主題，表示「此主題被提出來討論」的意思。

上班族加油站

在約定見面的信函中，最後應該要再提到預定見面的日期或地點，其他說法像是：I look forward to seeing you in Taipei on January 4.

 換你試試看!

1. 在你來訪的這段期間，我已經替你在 Belmont Suites 飯店安排好房間。

2. 我也為你三天的參訪規劃了行程。

3. 我想提醒你台北現在的天氣非常溫暖，所以你這趟旅程或許想要簡便的行李即可。

4. 希望你飛行愉快。

答案請參閱第 369 頁

中文翻譯

收件人：柯維斯
主旨：回覆：即將在台灣會面

親愛的維斯：

謝謝你確認日期，並寄給我即將到台北的行程細節。我非常期待這次的拜訪。

如果想到任何問題的話，我一定會讓你知道。

再次感謝這些資訊。一月四日我會在台北跟你見面。

謹上

喬

V. 電話安排會面一定要會這樣說

MP3 TRACK 68

訂票與訂房

A: **I just wanted to confirm that I have booked** my airline tickets to visit your office this summer.

我只是要確認我已經訂好這個夏天去你公司拜訪的機票了。

B: That's great to hear. We'll look forward to seeing you then.

很高興聽到這個消息。我們期待屆時見到你。

A: **I have made a reservation for you at** the Chelsea Grand Hotel.

我已經在切爾西大飯店幫你預訂了房間。

B: Thanks. I appreciate your assistance.

謝謝。我很感謝你的協助。

行程計劃

A: I will be in town for about three days.

我會在城裡待三天左右。

B: Great. **I have organized a complete itinerary for your trip.**

好的。我已經為你這次旅程規劃一個完整的行程。

提醒事項

A **Just a reminder that** the summer weather here can be very hot and humid.

只是要提醒，這裡夏天的天氣可能會非常炎熱潮溼。

B Thanks for the reminder. I'll be sure to bring the right clothes.

多謝提醒。我會帶些合適的衣服。

A **Please note that people here** dress pretty conservatively.

請留意這裡的人穿著都相當保守。

B That's good to know. Thanks.

我知道了。謝謝你。

預祝旅途愉快

A **Have a safe flight** and see you in Taipei.

祝你飛行平安，台北見囉。

B Thanks. Looking forward to it.

謝謝。我很期待。

Ⅵ. 如何用電話安排會面

海倫打電話給傑瑞確認他即將到高雄的行程安排及氣候等相關細節。

LISTENING 請聽 MP3 TRACK 69 ☐ SPEAKING 請跟著 MP3 唸唸看 ☐

Helen: Good morning. I'm calling for Jerry Smithers.

Jerry: Jerry here. What can I do for you?

Helen: Jerry, this is Helen Wu. I am just calling to confirm the specifics of your upcoming meeting with us here in Kaohsiung.

Jerry: Good thinking. I've been wanting to go over this with you.

Helen: Well, you're scheduled to be in Kaohsiung for five days. I have booked a suite for you at the Harbor Hotel. (訂房)

Jerry: Great. I enjoyed my stay there last time.

Helen: According to our itinerary, we have meetings scheduled at our office for most of your visit. (說明行程安排) But your final day will be free for sightseeing.

Jerry: Good. What will the weather be like when I visit?

Helen: It should be a bit hot, I'm afraid.

Jerry: That's OK. I'm still looking forward to meeting with you in Kaohsiung soon.

早安，我想找傑瑞・史密瑟斯。

我就是傑瑞。請問有什麼我可以效勞的地方？

傑瑞，我是吳海倫。我打來只是想確認你即將與我們在高雄會面的細節。

妳設想真是周到。我一直想和妳再確認這件事。

是這樣的，你預計在高雄待五天。我已經幫你在港口飯店訂好一間套房。

太好了。我上次住在那裡很喜歡。

根據我們的行程計畫，你來訪時大部分的時間都安排要在我們辦公室開會。不過最後一天可以自由觀光。

好的。我去拜訪時天氣會怎麼樣？

恐怕會有點熱。

沒關係。我還是期待能快點與妳在高雄碰面。

Try it! 換你試試看!

WRITING	請依提示寫出完整句子	☐
LISTENING	請聽 MP3 TRACK 70	☐
SPEAKING	請跟著 MP3 唸唸看	☐

1. Ⓐ Can you recommend a good hotel?

   Ⓑ ______________________________

   ____

   (reserve / suite / Royal Inn / length / stay)

   Ⓐ Thanks for your help.

2. Ⓐ I'd like to review with you the details of our upcoming meeting in Taichung.

   Ⓑ ______________________________

   ____

   (have / come up / itinerary / your visit)

3. Ⓐ What's the weather like in Taipei in February?

   Ⓑ ______________________________

   ____

   (get / quite chilly / rainy / February)

4. Ⓐ I will see you in Taipei next Monday.

   Ⓑ ______________________________

   ____

   (wish / smooth / journey)

   Ⓐ Thank you.

答案請參閱第 369 頁

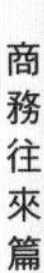

UNIT 11 取消會面

Canceling a Meeting

I. 取消會面一定要會的單字片語

MP3 TRACK 71

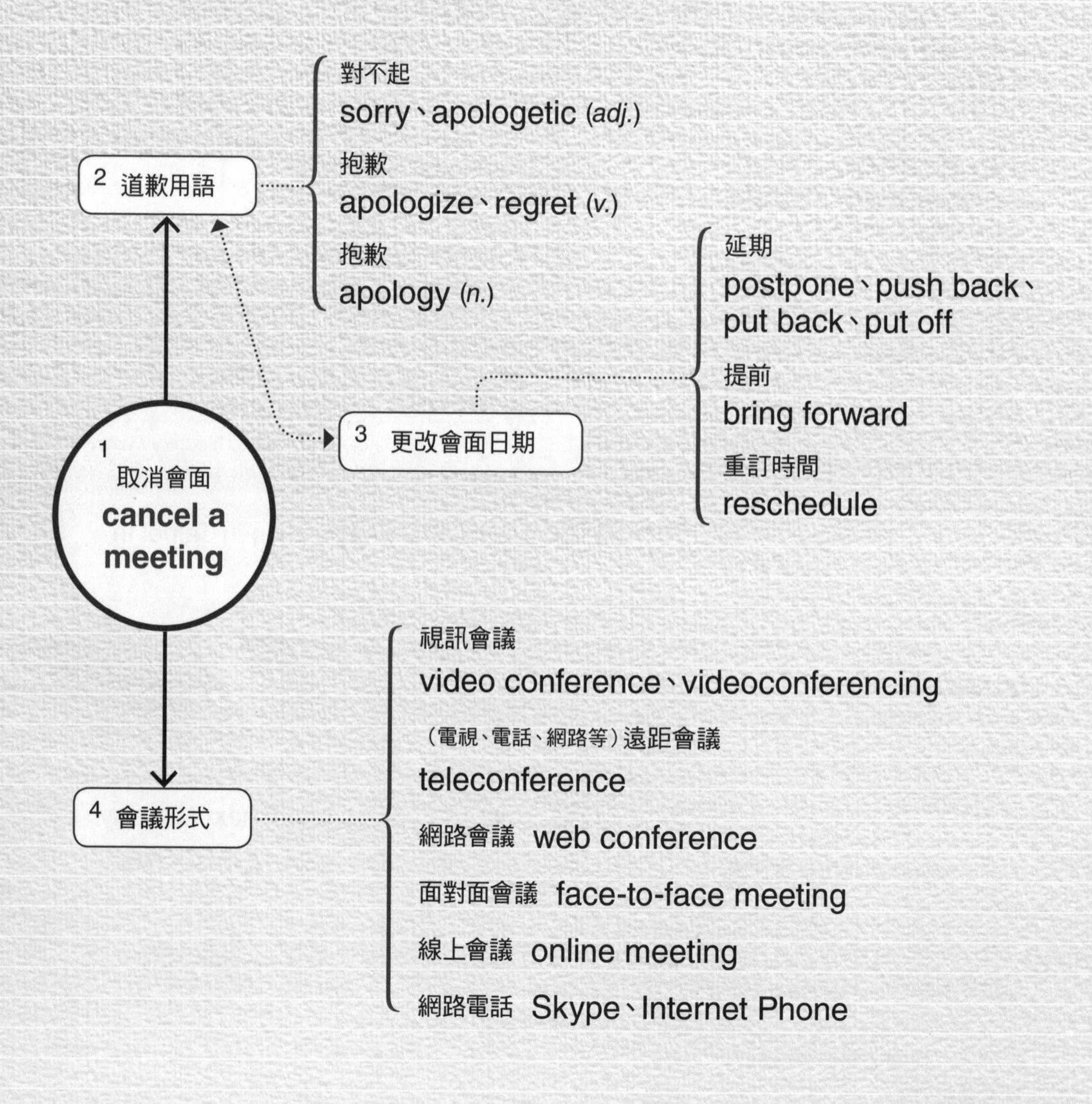

Ⅱ. 取消會面一定要會的句型

句型1 告知取消會面

I regret to say + that S. + V.（取消會面）

例 I regret to say that I must cancel my upcoming visit to Taiwan.
很抱歉，我必須取消即將去台灣的參訪。

句型2 說明取消原因

Unfortunately, sth **has come up** . . .

例 Unfortunately, an urgent matter has come up here in my company.
很遺憾，公司裡發生了緊急事件。

句型3 希望能改期

I would like to reschedule my visit for + 時間

例 I would like to reschedule my visit for this spring.
我希望能夠將參訪改到春天的時候。

句型4 其他補救方式

If you would still like to discuss sth, **perhaps we can arrange for** + 補救方法

例 If you would still like to discuss the matters we had scheduled, perhaps we can arrange for a teleconference via Skype.
如果你仍想討論我們原定的事項，或許我們可以安排 Skype 視訊會議。

Ⅲ. 如何用 e-mail 取消會面

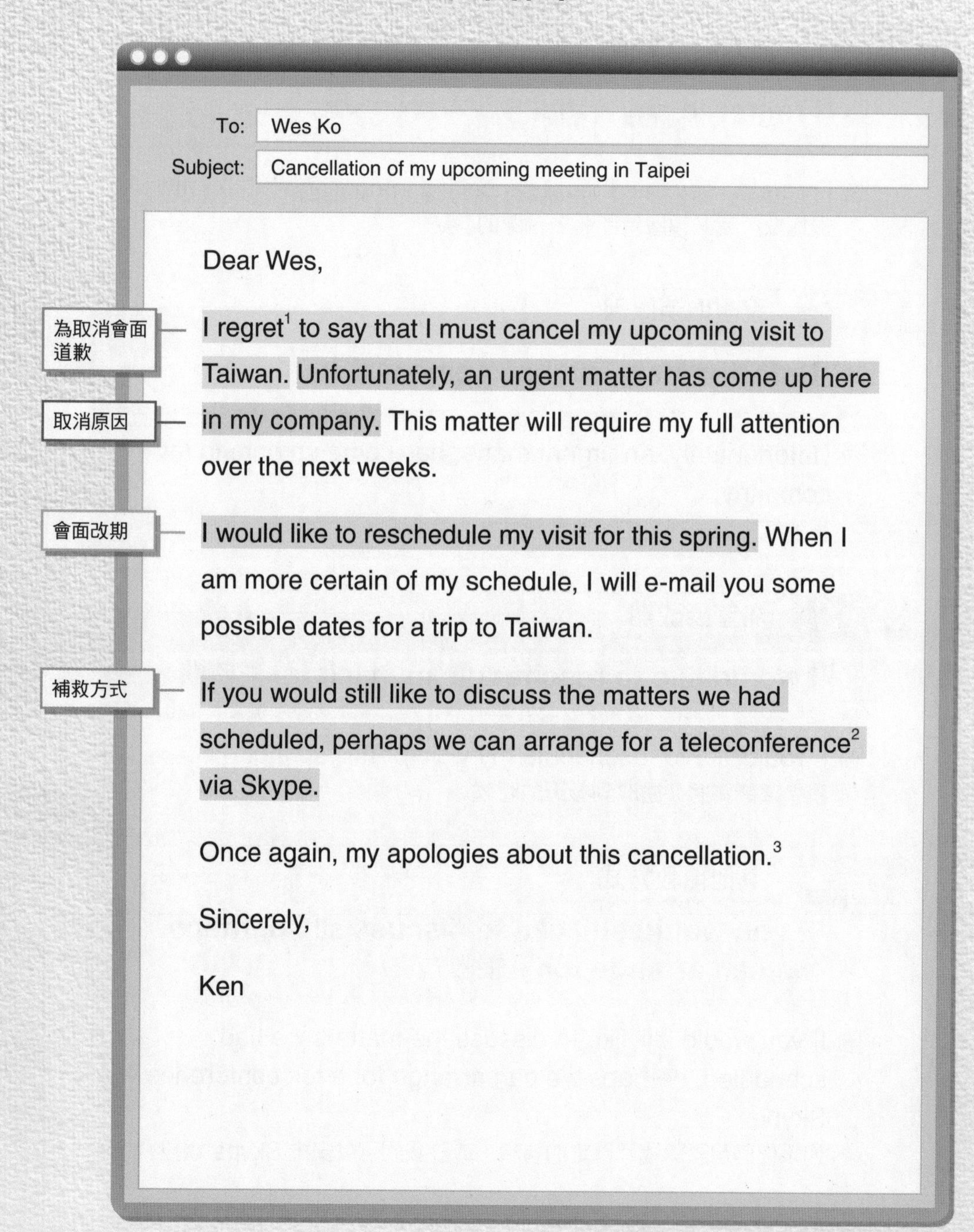

To: Wes Ko

Subject: Cancellation of my upcoming meeting in Taipei

Dear Wes,

為取消會面道歉 — I regret[1] to say that I must cancel my upcoming visit to Taiwan. 取消原因 — Unfortunately, an urgent matter has come up here in my company. This matter will require my full attention over the next weeks.

會面改期 — I would like to reschedule my visit for this spring. When I am more certain of my schedule, I will e-mail you some possible dates for a trip to Taiwan.

補救方式 — If you would still like to discuss the matters we had scheduled, perhaps we can arrange for a teleconference[2] via Skype.

Once again, my apologies about this cancellation.[3]

Sincerely,

Ken

Vocabulary & Phrases

1. regret [rɪˋgrɛt] *v.*
非常抱歉；遺憾

2. teleconference
[ˋtɛlɪˏkɑnfərəns] *n.* 透過電視、電話進行的遠距會議

3. cancellation [ˏkænsəˋleʃən] *n.*
取消（亦可拼成 cancelation）

中文翻譯

收件人：柯維斯
主旨：取消即將於台北的會面

親愛的維斯：

很抱歉，我必須取消即將去台灣的參訪。很遺憾，公司裡發生了緊急事件，之後幾個禮拜我需要專心處理這件事。

我希望能夠將參訪改到春天的時候。等我時間較為確定後，我會再把可能去台灣的日期寄電子郵件給你。

如果你仍想討論我們原定的事項，或許我們可以安排 Skype 視訊會議。

對於取消會議，我再次致上歉意。

謹上

肯恩

文化補給站

在商業往來中，視訊會議的使用率越來越高，因為它不僅能省下旅途往返的時間，還能保有面對面對談的特質，幫助你維持與客戶及業務夥伴之間的緊密關係。想要開一場成功的視訊會議，你可以留意下面的基本禮儀：

- 沒有說話的時候盡量讓你的系統保持靜音狀態，以免對方聽到你環境中的雜音。

- 眼睛看著鏡頭，動作自然，就好像你在跟房間裡的人說話一樣。

- 說話的速度稍微慢一點，因為麥克風可能會讓你的聲音難以辨識。

- 視訊會議的訊息接收通常會慢一些，所以繼續往下一個議題前，最好先稍待幾秒，看看對方是否有建議或其他問題要提出。

- 就像出席任何商務會議時一樣，請穿著合宜的服裝。避免穿著太過鮮豔的顏色或令人分心的圖案。

Ⅳ. 如何用 e-mail 回覆對方取消會面事宜

To: Ken Fields

Subject: Re: Cancellation of my upcoming meeting in Taipei

Dear Ken,

Thank you for your e-mail. I was sorry to hear that you have had to cancel your upcoming trip to Taiwan. However, I understand that you have urgent business to deal with[1] right now.

Once you know your schedule for the spring, we can work together to reschedule another time to meet here.

In the meantime,[2] we can continue to communicate by e-mail and phone.

Thank you once again for letting me know about this cancellation.

Sincerely,

Wes

Vocabulary & Phrases

1. deal with [dil] *v.* 處理
2. in the meantime [ˋminˏtaɪm] *phr.* 在這段時間

中文翻譯

收件人：肯恩・菲爾茲
主旨：回覆：取消即將於台北的會面

親愛的肯恩：

感謝你的來信。很遺憾聽到你必須取消即將來台之行。不過，我理解此刻你有緊急事務得處理。

一旦你知道今年春天的行程以後，我們可以一起重新安排其他在此地見面的時間。

在這段時間內，我們可以透過電子郵件和電話繼續聯絡。

再次感謝你告知我這趟行程取消了。

謹上

維斯

Try it! 換你試試看!

1. 很抱歉，我必須要取消下個月到香港的旅程。

2. 我家裡發生急事我必須要照料。

(照料 attend to)

3. 我想要將這次旅行計畫往後延，也許是三月。

(延後 postpone . . . to)

4. 如果你想要的話，我們可以安排視訊會議。

(安排、設置 set up)

答案請參閱第 369 頁

V. 電話取消會面一定要會這樣說

MP3 TRACK 72

為取消會面道歉

A **I'm sorry, but I'm calling to** let you know that I have to cancel our scheduled meeting.

很抱歉，我來電是要告訴你我必須取消預定的會議。

B That's too bad. I hope everything is all right.

太遺憾了。希望你一切安好。

A I just want to confirm that we are scheduled to meet on Monday.

我只是想確認我們預定在星期一見面的事。

B **I apologize, but I am going to have to call off that meeting.**

我很抱歉，我將得取消這次的見面。

說明理由

A **We've had a crisis in our factory**, and I need to stay here to take care of it.

我們工廠發生了危機，我得留在這裡處理。

B I understand. Please let me know if there's anything I can do to help.

我瞭解。如果有任何我可以效勞的地方，請告訴我。

A	Your e-mail mentioned that you need to cancel our meeting.	你寄來的電子郵件提到你得取消我們見面的事。
B	Yes. **The workers in our factory are going on strike.**	沒錯，我們工廠的工人正在進行罷工。

建議改期

A	**Let's try to reschedule our meeting for** sometime next week.	我們試試看能否將會面改到下週某個時間。
B	Sounds good. Let me know which dates will work best for you.	聽起來不錯。再讓我知道哪天對你最方便。

補救辦法

A	If you have difficulty traveling, I could come to meet you at your office.	如果你不方便來的話，我可以去你辦公室和你見面。
B	That would be great. Let's set up a time.	那太好了。我們訂個時間吧。

VI. 如何用電話取消會面

山姆因為公司出了一點狀況，必須打電話跟路薏絲取消預定的會面。

LISTENING 請聽 MP3 TRACK 73 □ | SPEAKING 請跟著 MP3 唸唸看 □

Sam: Hi there. I'm just calling to speak with Louise Gu.

Louise: You've reached Louise.

Sam: Hi Louise. This is Sam Worthington. I'm very sorry, but I'm calling to cancel our upcoming meeting.

為取消會面道歉

Louise: Oh, thanks for letting me know. Is everything OK?

Sam: Yes, I'm fine. Unfortunately, we've had a bit of a crisis in our office, and I need to be here to handle it.

取消原因

Louise: I understand. If there's anything I can do to help, please let me know.

Sam: Thanks, but we have everything under control. I'd actually like to try to reschedule our meeting for sometime next month.

會面改期

Louise: Sure. Just let me know what will work best for you.

Sam: Thanks. In the meantime, if you'd like to go through some numbers, we could Skype each other.

補救方式

Louise: That could be a good idea. I'm still looking forward to meeting with you in person next month, though.

上班族加油站

Skype 為即時通訊軟體，這個字在口語中也可以當成動詞來用。另外要注意正確的發音為 [skaɪp]，最後一個音節不要發成 [pi] 的音。

妳好，我想找顧路薏絲。

我就是路薏絲。

嗨，路薏絲。我是山姆 ‧ 華斯頓。真的很抱歉，我來電是要取消我們近日即將見面一事。

噢，謝謝你告訴我。一切都還好吧？

是的，我還好。可惜我們公司出了點狀況，我得在這裡處理。

我瞭解。如果有任何需要我幫忙的地方，請告訴我。

謝謝，不過現在事情都在掌握之中。其實，我是想能否另外安排下個月找個時間來見面。

沒問題。只要告訴我你何時最方便就可以了。

謝謝。在這段期間，如果妳想跟我討論一些數目的話，我們可以用 Skype 聯絡。

那是個好點子。不過我仍然期待下個月能見到你本人。

Try it! 換你試試看!

WRITING 請依提示寫出完整句子	☐
LISTENING 請聽 MP3 TRACK 74	☐
SPEAKING 請跟著 MP3 唸唸看	☐

1. Ⓐ Hi, Ken. What's the problem?

   Ⓑ ______________________________

   ____

   (sorry / call / cancel / upcoming visit / Taiwan)

   Ⓐ That's OK. Please let me know if you need any assistance.

2. Ⓐ I got your e-mail about canceling our meeting. I hope everything is OK.

   Ⓑ ______________________________

   ____

   (unfortunately / urgent matter / come up / company / deal with)

3. Ⓐ I understand that we can't meet next week. Any thoughts about rescheduling?

   Ⓑ ______________________________

   ____

   (could / reschedule / October)

4. Ⓐ I look forward to meeting you at a later date.

   Ⓑ ______________________________

   ____

   (if / discuss / matters / scheduled / perhaps / arrange / teleconference / Skype)

   Ⓐ That's a great idea. Let's set up a time to do so.

答案請參閱第 369 頁

UNIT

12 感謝招待

Expressing Gratitude for Hospitality

I. 感謝招待一定要會的單字片語

MP3 TRACK 75

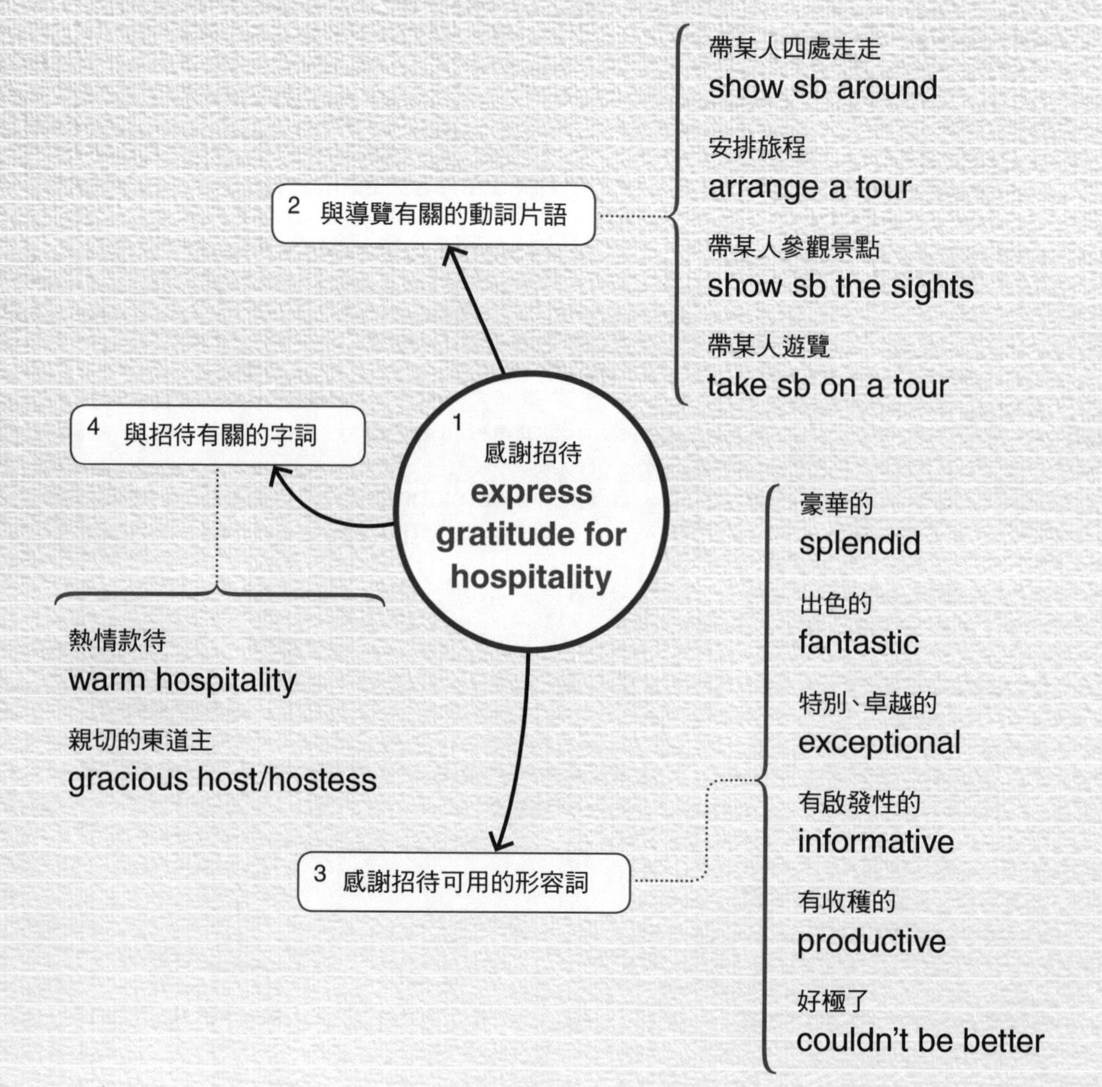

Ⅱ. 感謝招待一定要會的句型

句型 1 表達旅途愉快

. . . how much I enjoyed my recent trip to + 地點

例 Just a brief note to let you know how much I enjoyed my recent trip to Taiwan.
只是捎封短信讓你知道最近這趟台灣之旅讓我感到非常愉快。

句型 2 印象深刻或讓人懷念的事

It was a real treat + to V.

例 It was a real treat to visit the National Palace Museum.
去參觀故宮博物院是一件非常棒的事。

句型 3 感謝對方招待

It was nice of sb + to V.

例 It was nice of you to arrange a tour of the city for me.
承蒙你為我安排城市旅遊，不勝感激。

句型 4 期待有機會回報

On your next business trip to + 地點**, it would be my pleasure** + to V.

例 On your next business trip to Chicago, it would be my pleasure to take you around the city.
下次你到芝加哥出差時，希望有榮幸能帶你四處參觀。

Ⅲ. 如何用 e-mail 寫感謝招待函

To: Wes Ko

Subject: Recent Meeting in Taiwan

Dear Wes,

表達旅遊愉快

提到讓人懷念的事

感謝對方招待

Just a brief note to let you know how much I enjoyed my recent trip to Taiwan. It was very nice of you to arrange a tour of the city for me, and it was a real treat to visit the National Palace Museum.

I was pleased that we were able to come to an agreement regarding exclusive agency. We certainly look forward to working with you in entering this new and exciting market.

期待有機會回報

On your next business trip to Chicago, it would be my pleasure to take you around the city.

Thank you once again for all of your assistance during my trip.

Sincerely,

Ken

上班族加油站

感謝函中除了表達謝意外，還可提到令你覺得愉快或有收穫的事情，其他說法如下：

☑ **It was so exciting** to see the National Museum.

☑ **It was relaxing** spending the entire day on the beach.

☑ **What a treat it was / It was a real treat to** see a performance at the Opera House.

☑ **We especially enjoyed** the day we spent shopping / seeing the historic monuments.

中文翻譯

收件人：柯維斯
主旨：最近於台灣的會面

親愛的維斯：

只是捎封短信讓你知道最近這趟台灣之旅讓我感到非常愉快。承蒙你為我安排城市旅遊，不勝感激，並且去參觀故宮博物院是一件非常棒的事。

關於獨家代理一事，很高興我們能達成協議，我們真的非常期待與你們合作，進入這一塊生機蓬勃的市場。

下次你到芝加哥出差時，希望能有榮幸帶你四處參觀。

再次感謝你在我這趟旅程中所有的協助。

謹上

肯恩

文化補給站

感謝信該什麼時候寫呢？

寫感謝信最主要的目的就是要感謝對方的好意，就算只有短短幾句話也可以。不過，感謝信中不適合提到其他議題，如果你還要談其他公事的話，最好再另外寫一封信。

一般來說，只要你想表達謝意時都適合寫感謝信，因為大家都喜歡收到這樣窩心的信函，這一點並沒有語言或文化的差別。

如果是出差接受對方的招待，那麼感謝信最好在返家後一星期內寄出，不過嚴格來說，感謝信並沒有一定的禮儀規則，只是拖久了誠意就有點打折了。

Ⅳ. 如何用 e-mail 回覆感謝招待函

To: Ken Fields

Subject: Re: Recent Meeting in Taiwan

Dear Ken,

Thank you for your e-mail.

It was my pleasure to be your host during your trip to Taiwan. I'm glad that we had the chance to meet and come to an agreement.

I also enjoyed spending time showing you the local sights.

I look forward to seeing you again soon. Please remember, you are always welcome to visit us here in Taiwan.

Thank you once again for your visit.

Sincerely,

Wes

中文翻譯

收件人：肯恩．菲爾茲
主旨：回覆：最近於台灣的會面

親愛的肯恩：

謝謝你的來信。

在你來台期間能盡地主之誼是我的榮幸。我很高興我們有機會見面並達成協議。

帶你參觀本地景點的這段時間我也覺得很愉快。

期待能很快再與你見面。請記得，隨時都歡迎你到台灣來拜訪我們。

再次感謝你的到訪。

謹上

維斯

Try it! 換你試試看!

1. 我想要謝謝你讓我這趟台灣之旅圓滿成功。

2. 你人真好，帶我們在城裡四處參觀。

(四處參觀 take sb around)

3. 我們尤其開心看到了台北一〇一。

4. 下次你到芝加哥時我期待能回報你的款待。

(回報 reciprocate)

答案請參閱第 369 頁

商務往來篇

V. 用電話感謝招待一定要會這樣說

MP3 TRACK 76

表達此行愉快

A **I'm just calling to let you know how much I enjoyed my recent business trip** to Taiwan.

我打來只是想告訴你我最近這次去台灣出差非常愉快。

B That's great to hear. We enjoyed having you.

很高興聽到你這麼說。你來訪我們也很開心。

感謝對方招待

A **Thank you for being such gracious hosts.**

謝謝你的盛情款待。

B It was our pleasure. I'm glad that everything worked out.

這是我們的榮幸。我很開心一切都順利進行。

A **I appreciate all of your assistance** during my recent business trip.

我非常感謝在我最近出差期間你所提供的所有協助。

B I'm glad that I could be of help.

我很開心能幫上忙。

令人懷念的事情

A **Thank you again for taking me out for** such a memorable dinner. The food was exceptional.

再次感謝你帶我去吃令人難忘的晚餐，那裡的食物非常特別。

B That restaurant is famous for its food. I'm glad you liked it.

那間餐廳的菜色非常有名，我很高興你喜歡。

期待有機會回報

A You were such a great host. **I look forward to the opportunity to reciprocate your hospitality.**

你真的是一個很棒的東道主，我期盼有機會能回報你的熱情款待。

B And I look forward to visiting you and your company soon.

我也期待能很快去拜訪你和貴公司。

A Thanks for everything. **It would be my pleasure to take you around** the next time you visit St. Louis.

感謝你所做的每一件事。下次你來聖路易斯時，希望我有這個榮幸帶你到處看看。

B Oh, that's not really necessary, but thanks.

噢，不必這麼客氣，不過還是謝謝。

VI. 如何用電話感謝招待

伊莉莎．督柏森打電話給胡喬治，感謝他在她來台期間的招待。

LISTENING 請聽 MP3 TRACK 77 □　SPEAKING 請跟著 MP3 唸唸看 □

Eliza: Hello. I'm calling for George Hu.

George: Hello, this is George.

表達此行愉快

Eliza: Hi, George. This is Eliza Dobson. I'm just calling to let you know how much I enjoyed my recent business trip to Taiwan.

George: Oh, that's great to hear. It was a pleasure to have you visit.

感謝對方招待

Eliza: Well, you were a great host. Thanks for all your help in making my trip a success.

George: I'm glad that our negotiations went so smoothly.

提到印象深刻的事

Eliza: I think that's because we were able to meet face-to-face. I also enjoyed all of the great restaurants we went to.

George: We certainly had some memorable meals. I'm glad you enjoyed them.

期待有機會回報

Eliza: Well, on your next trip to America, I am going to return the favor.

George: Great. I look forward to it.

哈囉，我想找胡喬治。

哈囉，我就是喬治。

嗨，喬治。我是伊莉莎·督柏森。我打來只是要讓你知道最近到台灣出差的旅程讓我覺得很開心。

噢，真高興聽到妳這樣說。妳能來訪是我們的榮幸。

嗯，你真的是一個很棒的東道主。謝謝你所有的協助，讓我此行圓滿成功。

我很開心我們的協商順利進行。

我想那是因為我們有機會面對面談。我也很喜歡所有我們去的那些很棒的餐廳。

我們確實吃了一些令人懷念的餐點。我很高興妳喜歡。

那麼，你下次來美國時，我一定會回報你。

太好了，我非常期待。

Try it! 換你試試看!

WRITING 請依提示寫出完整句子	☐
LISTENING 請聽 MP3 TRACK 78	☐
SPEAKING 請跟著 MP3 唸唸看	☐

1. Ⓐ Hi, Sandy. Good to hear from you.

   Ⓑ Hi Tim. ______________________

   ____

   (calling / let / know / what / wonderful / trip)

   Ⓐ That's great to hear. It was a pleasure to have you.

2. Ⓐ I'm glad that I was able to be your host during your recent business trip.

   Ⓑ ______________________

   ____

   (appreciate / help / making / trip / success)

3. Ⓐ Thank you for taking the time to travel all the way to Taiwan to meet with us.

   Ⓑ ______________________

   ____

   (great / meet / in person / discuss / business)

4. Ⓐ It was my pleasure to show you around the city.

   Ⓑ Thank you. ______________________

   ____

   (please / remember / always / welcome / our guests)

   Ⓐ I look forward to visiting you and your company in America soon.

答案請參閱第 370 頁

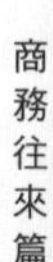

UNIT

13 請求代理權

Request for Agency

I. 請求代理權一定要會的單字片語

MP3 TRACK 79

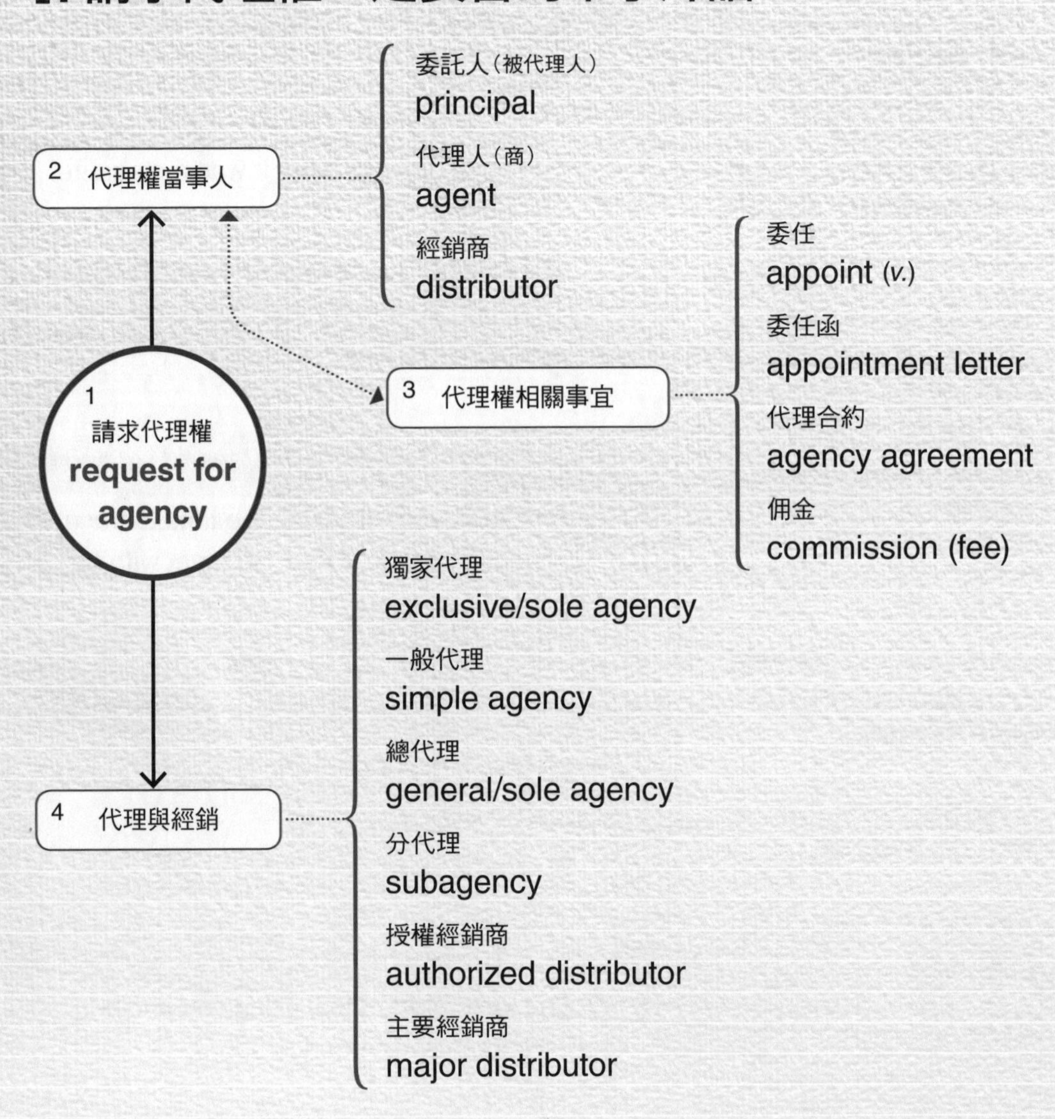

Ⅱ. 請求代理權一定要會的句型

句型 1 請求代理權

I am writing to express my interest in acquiring the rights of . . .

例 I am writing to express my interest in acquiring the rights of exclusive agency in Taiwan for the distribution, promotion, and sales of Jacob & Marsh's gourmet products.

我寫這封信是要表明對於取得雅各與馬爾斯精緻美食在台灣經銷、宣傳和販售方面獨家代理權的興趣。

句型 2 提出公司優勢

We have an influential presence in sth（產業、市場）

例 We also have an influential presence in the local food and wine community.

我們在本地的食品及酒類市場亦佔有一席之地。

句型 3 說明雙方合作的好處

To find out what benefits you can have by partnering with us . . .

例 To find out what benefits you can have by partnering with us, please contact us to set up a meeting to discuss this opportunity further.

想瞭解與我方合作能帶給你們什麼樣的好處，請與我們聯絡，以安排會議進一步討論這個機會。

Ⅲ. 如何用 e-mail 請求代理權

To: corporate_relations@jacobmarsh.com

Subject: Request for Exclusive Agency in Taiwan

> **上班族加油站**
> 這裡的 acquiring 也可以換成 gaining 或是 being granted。

Dear Sir/Madam,

開宗明義說明請求代理權

I am writing to express my interest in acquiring the rights of exclusive agency in Taiwan for the distribution, promotion, and sales of Jacob & Marsh's gourmet products.

說明公司的優勢

We, at Sage Enterprises, are an established business organization with over twelve years of experience in gourmet food import and distribution. We also have an influential presence in the local food and wine community.

As you may be aware, Taiwan is an expanding market for gourmet food products. Sage Enterprises will not only increase the presence of your brand here in Southeast Asia, but we will also compile market research information to be used in making forecasts.

提出雙方合作的好處

To find out what benefits you can have by partnering with us, please contact us to set up a meeting to discuss this opportunity further.

In the meantime, please don't hesitate to call or write to me if you have any questions.

Yours truly,

Jo Hsiao, Director

Sage Enterprises

中文翻譯

收件人：corporate_relations@
jacobmarsh.com
主旨：請求台灣獨家代理權

親愛的先生女士：

我寫這封信是要表明對於取得雅各與馬爾斯精緻美食在台灣經銷、宣傳和販售方面獨家代理權的興趣。

我們賽吉企業是擁有超過十二年食品進口、經銷經驗的老字號企業。我們在本地的食品及酒類市場亦佔有一席之地。

你可能會注意到，台灣是精緻美食產品擴張中的市場。賽吉企業不僅能提升你們品牌在東南亞的知名度，而且還能收集市場研究資訊來做預測。

想瞭解與我方合作能帶給你們什麼樣的好處，請與我們聯絡，以安排會議進一步討論這個機會。

這段時間，如果你有任何問題，請不要客氣，儘管打電話或寫信給我。

謹上

經理 蕭喬
賽吉企業

文化補給站

代理商好還是經銷商好？

選擇透過代理商（agent）或經銷商（distributor）進入國外市場有不同的好處。

代理商一般來說就是銷售代理商（sales agent），為國外廠商或供應商在海外市場的代理人，以收取佣金（commission）作為報酬的公司或個人，本身並不負盈虧之責。根據大多數的代理合約，代理商通常需負責整個行銷作業，包括建立品牌形象（brand image）、商品定價、文宣、銷售等，而廠商得要負責出貨（shipping）、文書作業（paperwork）以及其他貿易相關的物流（logistics），廠商對代理商有較大的約束力。

經銷商則是跟廠商進貨，然後到海外市場販售以賺取利差，至於銷售結果需自負盈虧。與經銷商合作的好處是省掉上述與選擇代理商所需負責的等等作業，然而經銷商可能會要求非常大的折扣和極為優渥的信貸條件（credit terms）。

這裡有幾個名詞你可以瞭解一下，不過雙方的權利義務還是要依照合約來決定。

sole/general agency	總代理
sole/general agent	總代理商
exclusive/sole ageny	獨家代理
exclusive/sole agent	獨家代理商
sole distributor, chief dealer	總經銷商

IV. 如何用 e-mail 回覆請求代理權

To: Jo Hsiao

Subject: Re: Request for Exclusive Agency in Taiwan

Dear Ms. Hsiao,

Thank you for getting in touch with us, and for expressing your interest in acquiring exclusive agency from Jacob & Marsh for Southeast Asia.

We are indeed interested in your proposal. Over the past few years, our products have sold very well in Europe, and we are now evaluating[1] the prospect[2] of expanding our market presence into Southeast Asia. However, since you are requesting to act as our exclusive agent, I do need some time to formulate[3] the terms with my associates first. I will therefore get back to you again, once we have outlined[4] an agreement to discuss with you.

Thank you.

Kind regards,

John Smith
Marketing Executive
Jacob & Marsh, Co.
126, Abercrombie Road
Santa Monica, CA 90405

Vocabulary & Phrases

1. evaluate [ɪˋvæljə͵wet] *v.* 評估
2. prospect [ˋprɑ͵spɛkt] *n.* （成功的）可能性
3. formulate [ˋfɔrmjə͵let] *v.* 制訂
4. outline [ˋaut͵laɪn] *v.* 制定大綱

中文翻譯

收件人：蕭喬
主旨：回覆：請求台灣的獨家代理權

親愛的蕭小姐：

感謝妳與我們聯絡，也感謝妳對於取得雅各與馬爾斯在東南亞的獨家代理權有興趣。

我們確實對妳的提議感興趣。在過去幾年，我們的產品在歐洲銷售得非常好，現在我們正在評估將市場擴展到東南亞的可能性。然而，由於妳要求作為我們的獨家代理商，我需要一些時間先和合夥人制定條款。一旦我們將與妳討論的合約要點概述出來後，我會再與妳聯繫。

謝謝妳。

誠摯的祝福

約翰·史密斯
行銷經理
雅各與馬爾斯公司
90405 加州聖塔莫尼卡市
亞伯克隆比路一百二十六號

Try it! 換你試試看!

1. 我寫信是要表達我有興趣在台灣經銷與販售貴公司的產品。

____

____

____

2. 我們 TLC 經銷商顯示了在過去八年來持續的成長，而且目前有五間分公司遍及東南亞。

____

____

____

(顯示 demonstrate　分公司 branch office)

3. 我們不僅能增加你們整體銷售額，而且能帶給你們只有我們才能提供的許多其他優勢。

____

____

____

4. 欲進一步討論這個成長的機會，請不要客氣，儘管和我聯繫。

____

____

____

答案請參閱第 370 頁

V. 用電話請求代理權一定要會這樣說

MP3 TRACK 80

請求代理權

A **I'm calling to discuss the possibility of becoming your company's agent** in Malaysia.

我打電話是要討論成為你們公司馬來西亞代理商的可能性。

B I'm sorry, but we already have a sole agent in that region, so we cannot enter into business relations with another company.

我很抱歉，我們那個地區已經有總代理商了，所以我們無法和其他公司建立業務關係。

委任代理

A Have there been any developments regarding the possibility of exclusive agency?

關於獨家代理的可能性有沒有任何進展？

B Yes. **We are pleased to confirm the appointment of your company as** our exclusive agent in the region we discussed.

有的。我們很高興確認由你們公司來擔任我們所討論地區的獨家代理商。

說明公司優勢

A **Our company has been in business for** more than thirty years.

我們公司從事買賣已經超過三十年了。

B This is one reason we've been looking to work with you.

這是我們一直都想與你們合作的一個原因。

合作所能帶來的好處

A With OPA, Co. as your exclusive agent, **your company will be one of the biggest players on** the Asian market.

有 OPA 公司作為你們的獨家代理商，貴公司將成為亞洲市場最具主導地位的公司之一。

B It is an exciting opportunity for us.

對我們來說是個令人振奮的機會。

player 在此指的是商場具有舉足輕重地位的公司。

提議付諸行動

A We invite you to visit us so we can **further discuss this mutually beneficial business relationship**.

我們邀請你來拜訪我們，這樣我們才能進一步討論互惠的業務關係。

B Thank you. I will get back to you in the next few days.

謝謝你。過幾天後我會跟你聯繫。

VI. 如何用電話請求代理權

凱倫打電話給弗萊德確認是否能擔任他們的獨家代理商，並約時間討論進一步細節。

LISTENING 請聽 MP3 TRACK 81 ☐	SPEAKING 請跟著 MP3 唸唸看 ☐

Karen: Hi. Is Fred Dobson there?

Fred: Yes, this is Fred. How may I help you?

Karen: This is Karen Lee from Lucky Trading Co. I'm calling to confirm your interest in granting us exclusive agency in Taiwan.

請求代理權

Fred: That's right. I sent you an e-mail about it last week.

Karen: Right. As you know, we are one of the most well-established trading firms in the leather goods sector in Taiwan.

說明公司的優勢

Fred: Yes, I am aware of that, and I think that Taiwan may be a good market for our brand.

Karen: As your exclusive agent, we would certainly make your interests our very own.

提出雙方合作的好處

Fred: That would certainly give us quite an advantage.

Karen: We would offer many other benefits to your company as well.

Fred: Well, Karen, I look forward to setting up a meeting with you to iron out the details of an agreement.

Karen: Wonderful. If you'd like, you could e-mail us a copy of the appointment letter and agency agreement to review first.

Fred: Sounds good. Talk to you soon.

嗨。請問弗萊德・督柏森在嗎？

是的，我就是弗萊德。有什麼可以為妳效勞的嗎？

我是幸運貿易公司的李凱倫。我來電是要確認你是否有興趣允許我們成為台灣的獨家代理。

沒錯。上個星期我有寄信跟妳提到此事。

沒錯。如你所知，我們是台灣皮件業最有口碑的貿易公司之一。

是的，我瞭解，也認為台灣對我們品牌來說可能是不錯的市場。

身為你們的獨家代理商，我們絕對會將你們的利益視為自己的利益。

那確實會帶給我們相當的好處。

我們也會提供貴公司許多其他助益。

那麼，凱倫，我期待能和妳安排會面一起來討論合約細節。

太棒了。如果你想的話，你可以用電子郵件將委任函以及代理合約先寄給我們看看。

聽起來不錯。那儘快再跟妳聯絡。

Try it! 換你試試看!

WRITING	請依提示寫出完整句子	☐
LISTENING	請聽 MP3 TRACK 82	☐
SPEAKING	請跟著 MP3 唸唸看	☐

1. Ⓐ Hi, Emily. What did you want to talk to me about?

   Ⓑ ______________________________

   ____

   (call / discuss / possibility / acquire / exclusive agency / your company)

2. Ⓐ Can you give me some more details about your company?

   Ⓑ ______________________________

   ____

   (we / in business / over / seven years)

3. Ⓐ What are some of the benefits of having your company as our company's exclusive agent?

   Ⓑ ______________________________

   ____

   (with us as / exclusive agent / enjoy / benefit / thorough knowledge / local market)

4. Ⓐ Why should we choose you to be our exclusive agents?

   Ⓑ ______________________________

   ____

   (exclusive agency / certainly / offers / advantages / company)

答案請參閱第 370 頁

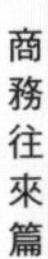

UNIT 14 | 請求著作權許可
Request for Copyright Permission

I. 請求著作權許可一定要會的單字片語

MP3 TRACK 83

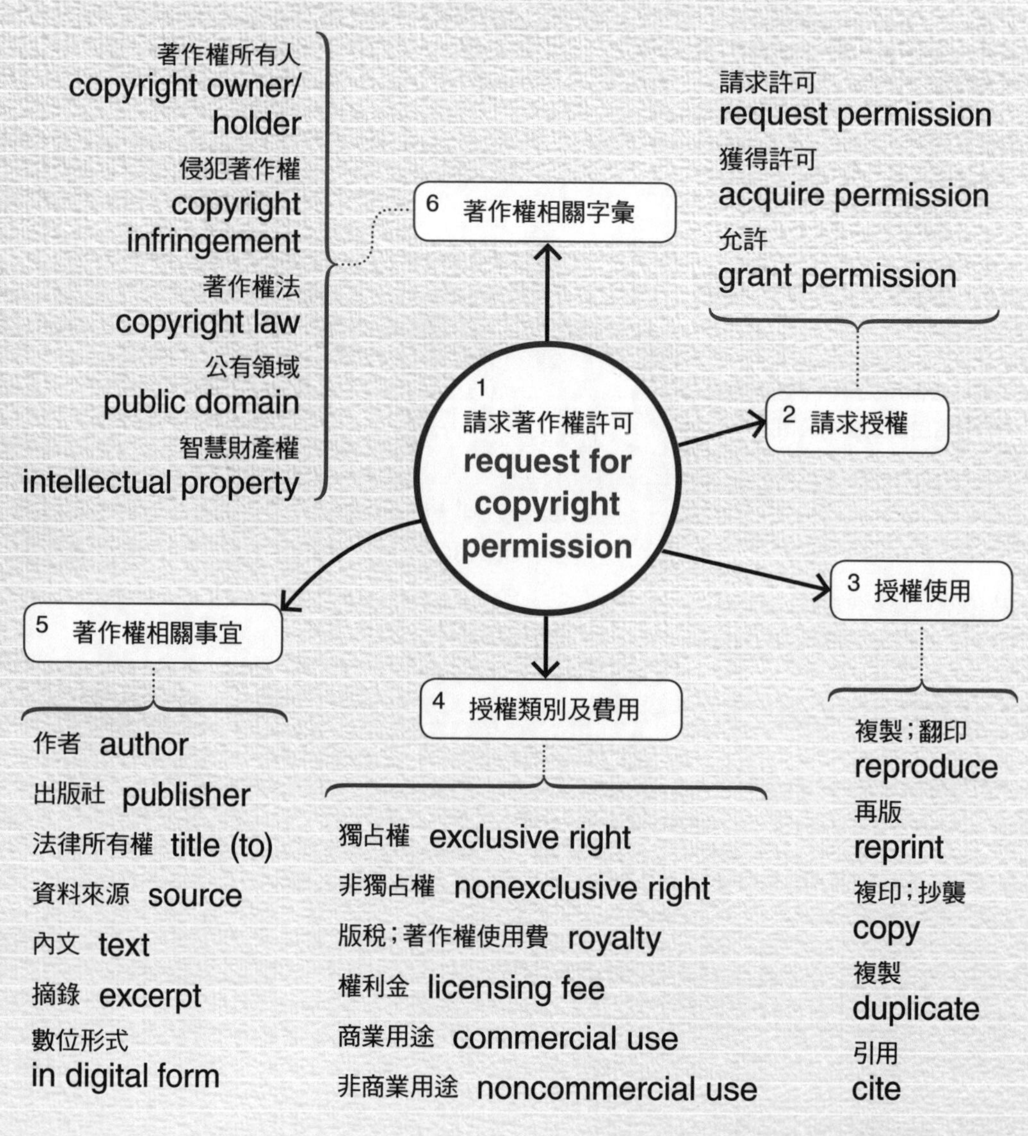

Ⅱ. 請求著作權許可一定要會的句型

句型 1 請求授權許可

I am writing to request permission to reprint sth.

例 I am writing to request permission to reprint the following article in the March 2012 issue of our magazine *Biz Buzz*:
我寫信是要請求允許下列的文章能在我們二〇一二年三月號的 *Biz Buzz* 雜誌上轉載：

句型 2 確認是否付費

Please let me know if you would require + 費用

例 Please let me know if you would require a flat licensing fee or royalties for permission to reprint.
請讓我知道允許複印的話是否會收取固定的權利金或版稅。

flat 在此表示「均一的；固定的」，與 fixed 意思相同

句型 3 確保權利不會受損

S. **will make sure to attribute the work correctly.**

例 I will, of course, make sure to attribute the work correctly.
當然，我一定會確保這篇文章的所有權無誤。

句型 4 確認著作權所有人

If you do not control copyright to . . ., I would appreciate . . .

例 If you do not control copyright to the requested material, I would appreciate any information you can provide about whom I should contact.
如果我們請求的內容你沒有著作權的話，若你能提供應該聯繫對象的資訊，我將感激不盡。

Ⅲ. 如何用 e-mail 請求著作權許可

To: Kelly Dobbs

Subject: Request for Copyright Permission

Dear Ms. Dobbs,

請求授權許可

I am writing to request permission to reprint the following article in the March 2012 issue of our magazine *Biz Buzz*:

Author: Seth Myers
Title: Getting out of Debt Wisdom
Source: http://www.resource-management.com.nz/articles/12012011.htm

Biz Buzz, published by Rover Media, is a print journal with a monthly distribution of over 150,000 copies throughout Taiwan via newsstands and subscriptions. The magazine investigates current issues in business and asset management.

確認是否要付費

Please let me know if you would require a flat licensing fee or royalties for permission to reprint. Any further details about payment are also welcome.

確保權利不會受損

I will, of course, make sure to attribute the work correctly.

確認著作權所有人

If you do not control copyright to the requested material, I would appreciate any information you can provide about whom I should contact.

I look forward to hearing from you.

Regards,

Carl Zhang, Associate Editor
Biz Buzz

中文翻譯

收件人：凱莉・都伯斯
主旨：請求著作權許可

親愛的都伯斯小姐：

我寫信是要請求允許下列的文章能在我們二〇一二年三月號的 *Biz Buzz* 雜誌上轉載：

作者：塞斯· 梅爾斯
標題：遠離債務的智慧
來源：http://www.resource-management.com.nz/articles/12012011.htm

Biz Buzz 是由洛福媒體所出版，是一本每月在台灣透過書報攤及訂閱發行量超過十五萬份的平面出版期刊。本雜誌針對目前企業和資產管理方面的議題進行調查。

請讓我知道允許複印的話是否會收取固定的權利金或版稅。任何關於費用的進一步細節也歡迎妳說明。

當然，我一定會確保這篇文章的所有權無誤。

如果我們請求的內容妳沒有著作權的話，若妳能提供應該聯繫對象的資訊，我將感激不盡。

敬候佳音。

謹上

副編輯　張卡爾
Biz Buzz

文化補給站

以下為取得著作權的步驟：

1. 確認是否有著作權

以美國來說，一九二二年以後首次出版的作品皆受著作權法的保護。然而複製內容只佔很小部分或作為教育用途時，則未經授權使用亦是合法的。

2. 確認著作權所有人

這點可能很簡單也可能很複雜。著作權所有人經年累月可能經過多次轉手，有時一件作品可能同時為許多個人或團體共同擁有。

3. 確認需要的權利

先確認所請求的是哪些權利。由於可能會包含一些費用，你不會想多付不必要的費用。這些內容的使用目的為何？是要取得獨家授權還是非獨家授權？授權許可的條款為何（如使用期限）？使用權是否會撤銷？是否只限定在某些地區使用？

4. 協商是否支付費用

有時你所摘錄的內容並不多，或作為教育、非營利用途而推廣複製作品的話，創作者並不會向你收取費用。另外有些情況是有些名不見經傳的藝術家或音樂人可能會為了增加曝光而暫時不收費，除非他們的作品開始獲利才會收費。

5. 白紙黑字寫下來

為了保護你的權利，避免口頭取得授權，最好要白紙黑字寫下來。

Ⅳ. 如何用 e-mail 回覆是否授權

To: Carl Zhang

Subject: Re: Request for Copyright Permission

Dear Mr. Zhang,

Thank you for your e-mail. We are delighted to learn that you would like to use one of our Web articles in your magazine. For one-time use, we normally charge a one-time flat fee of US$250. Please attribute the article to the author and cite the source where the article was originally published.

I have attached our standard copyright permission form. Please fill this out and return it to us at your earliest convenience.

Thank you again for your interest in our material.

Sincerely,

Kelly Dobbs,
Resource Management, Inc.

中文翻譯

收件人：張卡爾
主旨：回覆：請求著作權許可

親愛的張先生：

感謝你的來信。我們很高興得知你們想在雜誌裡使用我們網站上的一篇文章。對於一次性的使用，我們通常收取單次兩百五十美元的固定費用。請將這篇文章的所有權標示為作者，並註明文章的原始出處。

我已附上我們的標準著作權許可表格。請將表格填好，在你方便時儘快回傳給我們。

再次感謝你對我們的資料感興趣。

謹上

凱莉．都伯斯
資源管理公司

Try it! 換你試試看!

1. 我寫信是要請求同意下列照片能使用在我的文章中。

2. 我們願意為允許複印此篇文章而支付小額權利金。

3. 請讓我知道你想要我們如何標示文章的所有權。

4. 如果你並非著作權所有人，要是你可以給我聯絡適當的人或公司的任何資訊，我將感激不盡。

答案請參閱第 370 頁

V. 電話請求授權許可一定要會這樣說

MP3 TRACK 84

請求使用許可

A **I'm calling to request permission to use** some of your copyrighted material.

我打電話是想要請求允許使用受你們版權保護的資料。

B That's possible. What material do you want to use?

那是可能的。你想要使用什麼資料？

A Have you had a chance to look at the content on our Web site?

你有沒有機會看一下我們網站上的內容？

B Yes, and we **are interested in getting your permission to use** some of it on our site.

有的，我們想要在網站上使用一些上面的內容，希望能得到你的允許。

費用收取方式

A **Will we need to pay to use this content?**

我們使用這些內容需要付費嗎？

B Yes. We charge a flat licensing fee for use of our content.

是的。使用我們的內容我們會收取固定的權利金。

著作權歸屬

A When you use this material, **please remember to attribute it properly.**

當你使用這些資料時，請記得要正確標示其所有權。

B We'll be sure to do so.

我們一定會照辦。

A **Be sure to state** our company's name **as the owner of this material.**

務必將我們公司名稱標示為資料的所有人。

B I got it. We'll make certain that this gets done.

我瞭解了。我們會確保有做到這一點。

確認著作權所有人

A **Do you own the copyright to this material?**

你擁有這個資料的著作權嗎？

B Yes, I am the sole copyright owner.

是的，我是著作權唯一所有人。

A I'm afraid we no longer hold the copyright to these articles.

恐怕我們不再持有這些文章的著作權。

B **Could you tell me whom I should contact to get reprint permission?**

你可以告訴我應該聯繫誰才能取得複印的許可嗎？

Ⅵ. 打電話請求著作權許可

艾力克斯打電話詢問轉載某篇文章的著作權許可。

LISTENING 請聽 MP3 TRACK 85 | SPEAKING 請跟著 MP3 唸唸看

Alex: Hello. Alex Bai here. I'm calling for Sue Tripp.

Sue: This is Sue. What can I do for you?

請求使用許可

Alex: I work in a small publishing company in Taiwan. I'm calling to request permission to use some of your copyrighted content.

Sue: Thank you for your interest in our work. What exactly do you want to use?

Alex: Specifically, we are interested in using the article "Career Changes at Middle Age" It appeared in your *Business Topics* magazine, and it was written by Don George.

Sue: That's a very popular article.

確認是否要付費

Alex: When it comes to payment, what are your standard terms?

Sue: We typically charge a flat fee for one-time use of our articles.

Alex: OK. And do you have any guidelines for attributing articles?

確認著作權所有人

Sue: Just be sure to state the author's name and that the article was used with our permission.

你好。我是白艾力克斯。我想找蘇·崔普。

我是蘇。有什麼可以為你效勞的嗎？

我任職於台灣的一家小出版社。我來電是想請求允許使用一些受你們著作權保護的內容。

感謝你對我們的著作感興趣。你們確切想用的是什麼內容？

具體來說，我們有興趣使用〈中年轉職〉這篇文章。它出現在你們《商務議題》雜誌中，是由唐·喬治所撰寫。

那篇文章很受歡迎。

談到費用，你們的收費標準條款為何？

一次性的使用我們通常收取固定費用。

好的。文章所有權的歸屬你們有什麼要遵守的原則嗎？

只要確實標出作者姓名，及文章是經由我們允許使用即可。

Try it! 換你試試看!

WRITING	請依提示寫出完整句子	☐
LISTENING	請聽 MP3 TRACK 86	☐
SPEAKING	請跟著 MP3 唸唸看	☐

1. Ⓐ How may I help you?

   Ⓑ ______________________________

   ____

   (calling / request / permission / reproduce / excerpts / book / our / Web site)

   Ⓐ We'd be delighted to help you.

2. Ⓐ We require payment for the use of our copyrighted content.

   Ⓑ ______________________________

   ____

   (please / let / know / should / pay / flat fee / royalties / use)

3. Ⓐ Please be sure to indicate that this is copyrighted material.

   Ⓑ ______________________________

   ____

   (of course / sure / attribute / work / correctly)

4. Ⓐ I am the writer of these articles.

   Ⓑ ______________________________

   ____

   (you / also / copyright / owner / these / articles)

   Ⓐ No, the copyright belongs to XYZ publishers.

答案請參閱第 370 頁

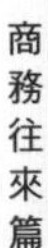

UNIT 15 介紹新的業務關係

Introducing a New Business Relationship

I. 介紹新的業務關係一定要會的單字片語

MP3 TRACK 87

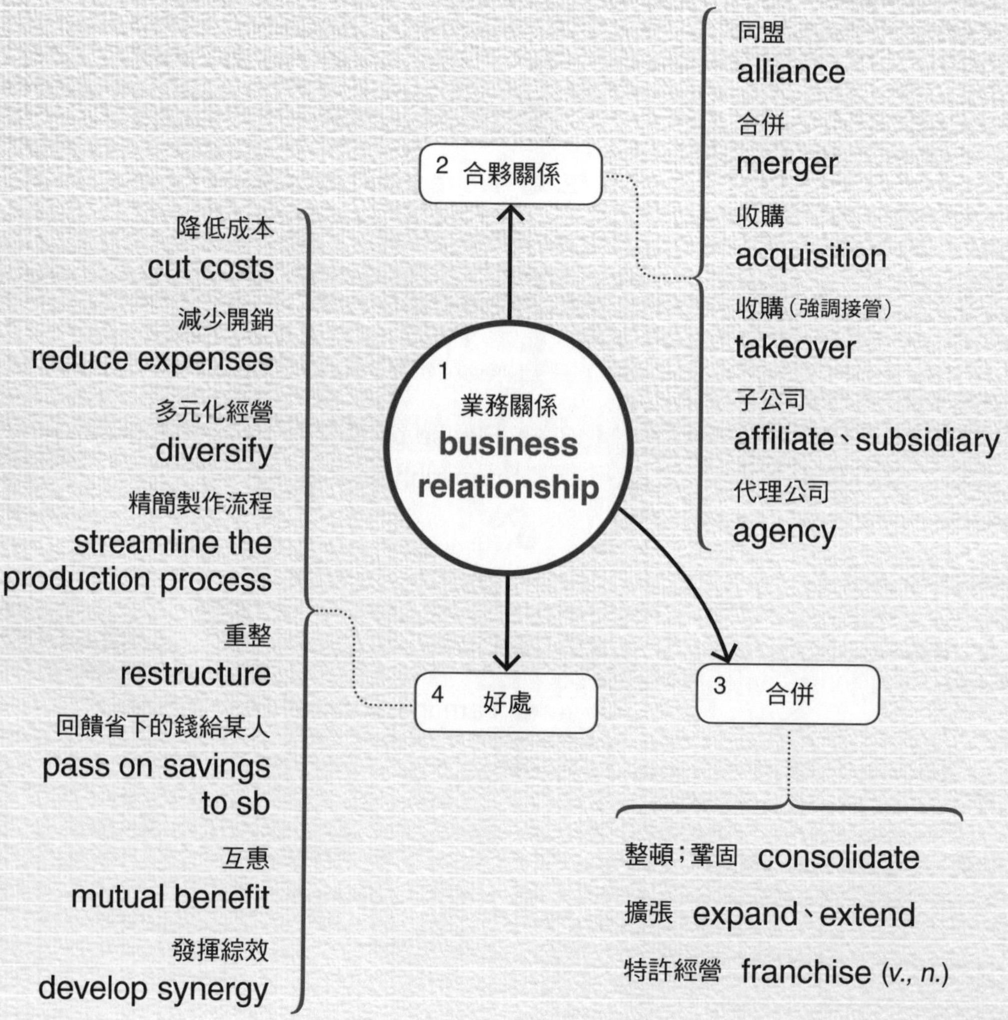

1 業務關係 **business relationship**

2 合夥關係
- 同盟 alliance
- 合併 merger
- 收購 acquisition
- 收購（強調接管） takeover
- 子公司 affiliate、subsidiary
- 代理公司 agency

3 合併
- 整頓；鞏固 consolidate
- 擴張 expand、extend
- 特許經營 franchise (*v., n.*)

4 好處
- 降低成本 cut costs
- 減少開銷 reduce expenses
- 多元化經營 diversify
- 精簡製作流程 streamline the production process
- 重整 restructure
- 回饋省下的錢給某人 pass on savings to sb
- 互惠 mutual benefit
- 發揮綜效 develop synergy

Ⅱ. 介紹新的業務關係一定要會的句型

句型 1　介紹新的業務關係

I am writing this brief e-mail to inform you of our company's new partnership with + 公司名稱

例 I am writing this brief e-mail to inform you of our company's new partnership with AccuTech.
我寫這封短信是要通知你我們公司和艾可科技新的合夥關係。

句型 2　未來的計畫

Our future plans include sth（計畫）

例 Our future plans include a complete merger of both companies.
我們未來的計畫包括兩家公司完全合併。

句型 3　帶來的好處

S. **will allow us** + to V.

例 This merger will allow us to increase our sales base and reduce our operating expenses.
合併會讓我們增加銷售基礎並降低我們的營運費用。

句型 4　對於貴公司的影響

As far as our business with your company goes, . . .

例 As far as our business with your company goes, this will remain unchanged.
至於和你們公司之間的生意往來將維持不變。

Ⅲ. 如何用 e-mail 介紹新的業務關係

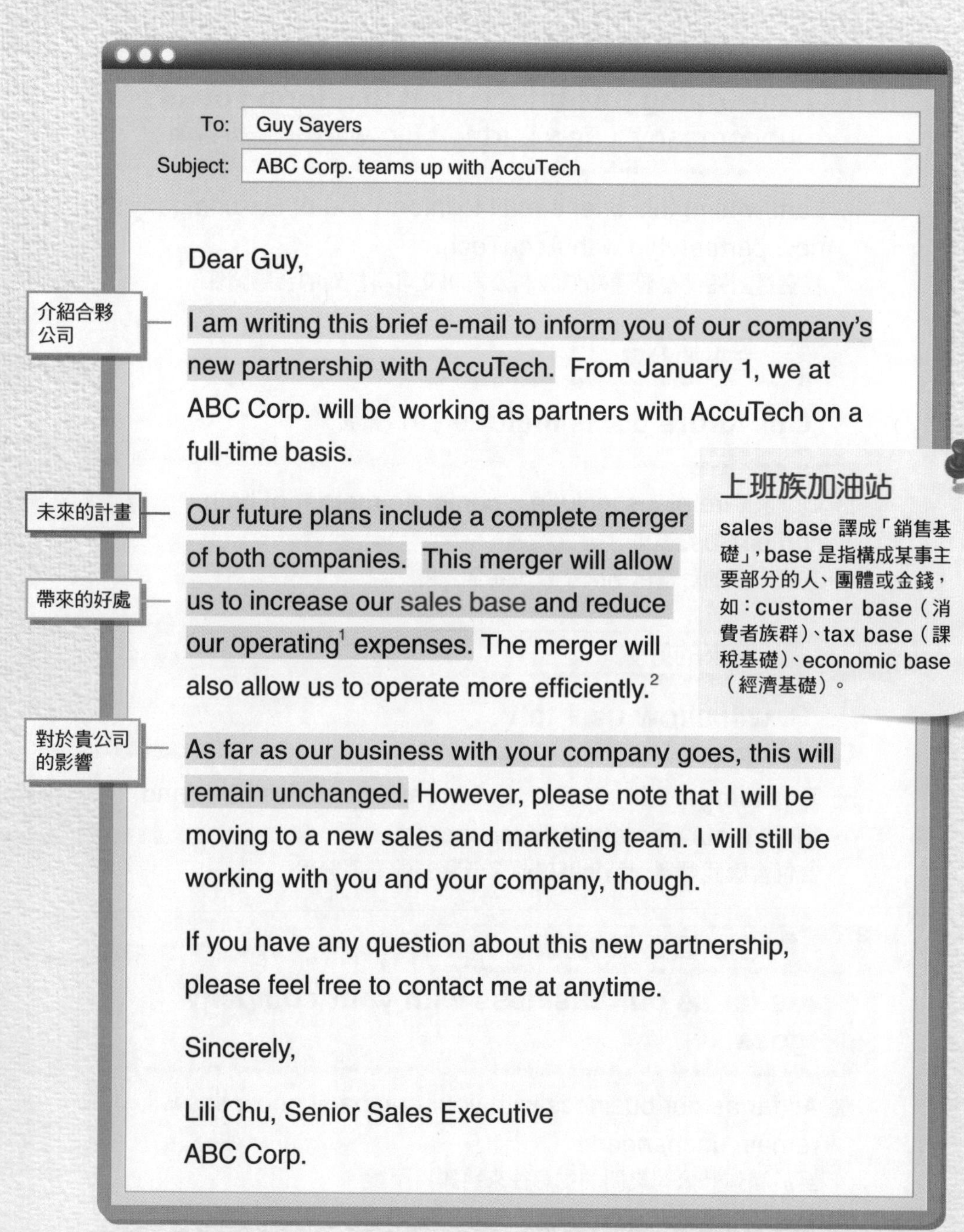

To: Guy Sayers

Subject: ABC Corp. teams up with AccuTech

Dear Guy,

介紹合夥公司 — I am writing this brief e-mail to inform you of our company's new partnership with AccuTech. From January 1, we at ABC Corp. will be working as partners with AccuTech on a full-time basis.

未來的計畫 — Our future plans include a complete merger of both companies. 帶來的好處 — This merger will allow us to increase our sales base and reduce our operating[1] expenses. The merger will also allow us to operate more efficiently.[2]

對於貴公司的影響 — As far as our business with your company goes, this will remain unchanged. However, please note that I will be moving to a new sales and marketing team. I will still be working with you and your company, though.

If you have any question about this new partnership, please feel free to contact me at anytime.

Sincerely,

Lili Chu, Senior Sales Executive
ABC Corp.

上班族加油站

sales base 譯成「銷售基礎」，base 是指構成某事主要部分的人、團體或金錢，如：customer base（消費者族群）、tax base（課稅基礎）、economic base（經濟基礎）。

Vocabulary & Phrases

1. operating [ˋɑpə͵retɪŋ] *adj.*
   營運的；運作的
2. efficiently [ɪˋfɪʃəntli] *adv.*
   效率高；有效率地

中文翻譯

收件人：蓋・賽爾斯
主旨：ABC 公司和艾可科技展開合作

親愛的蓋：

我寫這封短信是要通知你我們公司和艾可科技新的合夥關係。從一月一日開始，我們 ABC 公司和艾可科技會以正式合夥關係合作。

我們未來的計畫包括兩家公司完全合併。合併會讓我們增加銷售基礎並降低我們的營運費用，也會讓我們在營運上更有效率。

至於和你們公司之間的生意往來將維持不變。然而，請留意我將調到新的業務行銷團隊去。不過還是會持續和你及貴公司合作。

如果你對於新的合夥關係有任何問題，請隨時與我聯繫。

謹上

資深業務主任　邱莉莉
ABC 公司

文化補給站

據經濟合作發展組織（Organization for Economic Cooperation and Development）指出，儘管近年來全球經濟發展停滯，截至二〇一一年十月二十一國際企業間的合併（merger）和收購（acquisition）金額高達八千二百二十二億美元。大多數的國際投資從北美、西歐、中國（包括香港）等地區持續不斷釋出，為二〇一一年國際企業間合併受購活動的主要資金來源。美國及英國是企業併購活動的主要發生地，依序是中國、義大利及法國。

Ⅳ. 如何用 e-mail 回覆新業務關係的訊息

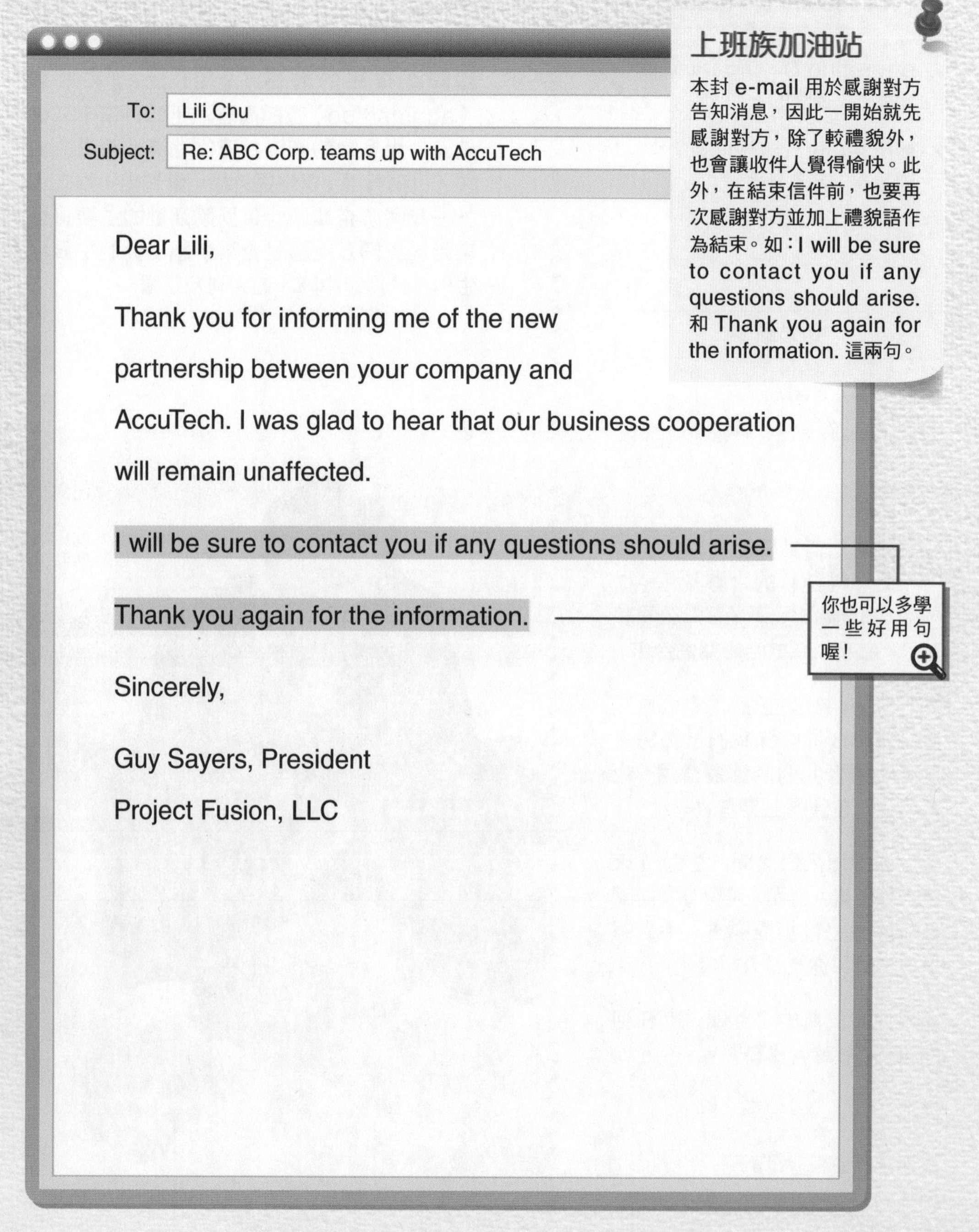

To: Lili Chu

Subject: Re: ABC Corp. teams up with AccuTech

Dear Lili,

Thank you for informing me of the new partnership between your company and AccuTech. I was glad to hear that our business cooperation will remain unaffected.

I will be sure to contact you if any questions should arise.

Thank you again for the information.

Sincerely,

Guy Sayers, President

Project Fusion, LLC

上班族加油站

本封 e-mail 用於感謝對方告知消息，因此一開始就先感謝對方，除了較禮貌外，也會讓收件人覺得愉快。此外，在結束信件前，也要再次感謝對方並加上禮貌語作為結束。如：I will be sure to contact you if any questions should arise. 和 Thank you again for the information. 這兩句。

中文翻譯

收件人：邱莉莉
主旨：回覆：ABC 公司和艾可科技展開合作

親愛的莉莉：

感謝妳通知我貴公司與艾可科技之間新的合夥關係。我很高興得知我們的業務合作不受影響。

如果有任何問題的話，我一定會與妳聯繫。

感謝妳的資訊。

謹上

總經理　蓋·賽爾斯
Project Fusion 有限公司

Try it! 換你試試看！

1. 我們要通知你 Peter Barnes 顧問公司和 Domino 公司之間新的合夥關係。

(顧問公司 Consultancy)

2. 兩家公司計畫在二〇一二年年底合併。

3. 藉由合併兩家公司，我們將能提供我們的客人更多的利益。

4. 我將到新的業務團隊，不過我仍會是你們主要的窗口。

(窗口、聯絡人 contact)

答案請參閱第 371 頁

V. 用電話介紹新業務關係一定要會這樣說

MP3 TRACK 88

告知新的合夥關係

A **I'm just calling to let you know that we are forming a new partnership with** Hills Co.

我打電話是要告訴你我們將與希爾斯公司建立新的合夥關係。

B Congratulations, and thanks for the information.

恭喜，也謝謝你的資訊 。

未來計畫

A **We're going to cooperate with** our new partner to enter the North American market.

我們即將與新的夥伴合作進入北美市場。

B With this new partnership, expanding to the North American market shouldn't be a problem.

有這個新的夥伴，要擴展北美市場應該不是問題。

A Because of the merger, **we will consolidate our operations.**

因為合併的關係，我們將整頓營運。

B Please keep me informed of any major changes.

任何重大變動請一定要通知我。

keep me informed of sth 是表示「隨時讓我知道某事」的意思，為固定用法。

好處

A | With this new partnership, **we'll be able to offer you an even wider range of products.**

有了新的合夥關係，我們就能提供你範圍更廣的產品。

B | Great. That will give us more choices.

太好了。那我們就有更多的選擇了。

業務變動

A | Once the partnership is complete, **I will be working in a new department.**

一旦合夥關係完成，我就會在新的部門上班。

B | Please send me your new contact information, when you get a chance.

你有空的時候請把新的聯絡資訊寄給我。

A | After the new merger, **our cooperation with you will continue as it always has in the past.**

新的合併之後，和你們的合作關係還是會和以前一樣持續下去。

B | It's good to know that some things won't change.

知道有些事情沒有改變是好的。

VI. 用電話介紹新的業務關係

李凱文打電話告知蕾・凱立他們即將與艾爾發系統整合公司建立合夥關係。

LISTENING 請聽 MP3 TRACK 89 ☐ | SPEAKING 請跟著 MP3 唸唸看 ☐

Kevin: Hi. This is Kevin Lee. Is Rae Kelley there, please?

Rae: Rae speaking. How are you this morning, Kevin?

介紹合夥公司

Kevin: Just fine. I'm just calling to let you know that our company will be forming a new partnership with Alpha Systems.

Rae: Is that right? Thanks for letting me know.

Kevin: This partnership will take place over the next few months.

Rae: Will there be any changes to the way we cooperate?

業務變動

Kevin: Not at all. We'll still work together in the same way, though I will be moving to a new department.

Rae: Be sure to send me your new contact info.

帶來的好處

Kevin: I will. This new partnership will make us more efficient, and we plan to pass these savings on to valuable customers like you.

Rae: That's always great to hear. Thanks again for the update.

嗨。我是李凱文。請問蕾·凱立在嗎?

我就是蕾。凱文,今天早上好嗎?

還可以。我打電話只是要告訴妳我們公司即將和艾爾發系統整合公司建立新的合夥關係。

是真的嗎?謝謝你讓我知道。

合夥關係會在幾個月後生效。

我們的合作模式會有任何改變嗎?

一點也不會。我們的合作方式還是一樣,不過我會到另一個部門去。

務必要將你新的聯絡資訊寄給我。

我會的。這次新的合作關係會讓我們更有效率,我們打算將省下來的花費回饋給像妳這樣的重要客戶。

聽到這種事總是感覺很好。再次感謝你最新的消息。

Try it! 換你試試看!

WRITING 請依提示寫出完整句子	☐
LISTENING 請聽 MP3 TRACK 90	☐
SPEAKING 請跟著 MP3 唸唸看	☐

1. Ⓐ What's the news, Jack?

   Ⓑ ______________________________

   ____

   (we / would like / inform / upcoming / merger / Venture Intelligence Company)

   Ⓐ Thanks for letting me know about this.

2. Ⓐ Can you tell us some more about your company's future plans?

   Ⓑ ______________________________

   ____

   (newly / merged / company / named / Apex Capital Ventures)

3. Ⓐ What are some of the benefits of this merger?

   Ⓑ ______________________________

   ____

   (merger / make / company / stronger / more competitive / market)

4. Ⓐ After the merger, will our companies continue to work together in the same way?

   Ⓑ ______________________________

   ____

   (your status / valued / partner / remain / unchanged)

答案請參閱第 371 頁

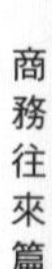

貿易活動篇

本篇章中介紹貿易往來最常見的狀況，包括詢問產品、價格、議價、下單、出貨、催款等貿易活動的流程，無論你是買方或賣方，我們提供你各種商務電子書信或電話英語的實用應對技巧。讓你在全球化的國際貿易中，輕鬆上手。

UNIT 16 詢問產品

Product Inquiry

I. 詢問產品一定要會的單字片語

MP3 TRACK 91

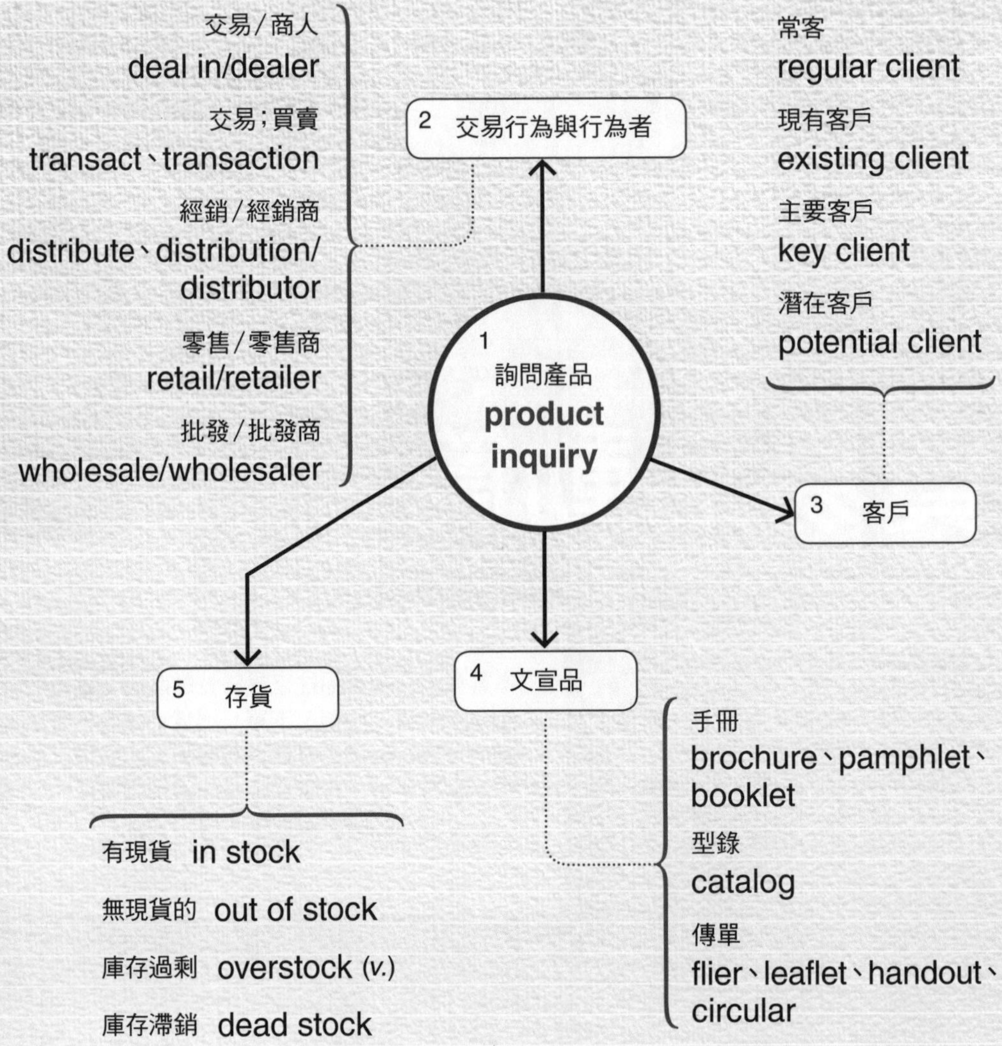

1 詢問產品 product inquiry

2 交易行為與行為者

交易／商人
deal in/dealer

交易；買賣
transact、transaction

經銷／經銷商
distribute、distribution/distributor

零售／零售商
retail/retailer

批發／批發商
wholesale/wholesaler

3 客戶

常客
regular client

現有客戶
existing client

主要客戶
key client

潛在客戶
potential client

4 文宣品

手冊
brochure、pamphlet、booklet

型錄
catalog

傳單
flier、leaflet、handout、circular

5 存貨

有現貨 in stock

無現貨的 out of stock

庫存過剩 overstock (*v.*)

庫存滯銷 dead stock

Ⅱ. 詢問產品一定要會的句型

句型 1 說明如何取得聯絡資訊

I received your contact information from sb.

例 I received your contact information from Christopher Holden, whom I believe you are familiar with.
我從克里斯多福．荷登那裡取得你的聯絡資訊，我想你應該跟他很熟。

句型 2 表達對產品感興趣

I am particularly impressed with sth（商品名稱）

例 I am particularly impressed with your Stack-and-Pack'em storage containers.
我對你們的堆疊式收納箱印象特別深刻。

句型 3 提出顧慮

My only concern is sth.

例 My only concern is availability.
我唯一擔心的是有沒有現貨。

availability 是指所要的物品可以取得或找到，在此用來表示有存貨可以使用的意思。

句型 4 索取資料

Could you please send me sth?

例 Could you please send me a comprehensive price list?
你可以寄給我一份綜合價目表嗎？

Ⅲ. 如何用 e-mail 詢問產品

Date: February 12, 2012

To: Yolinda Phelps

From: Sadams@deltainteriors.com

Subject: Product Inquiry: Stack and Pack'em Containers

Dear Ms. Phelps,

My name is Alice Chen. I own a chain of stores in Taiwan called Delta Interiors.[1] We specialize in practical interior solutions. I received your contact information from Christopher Holden, whom I believe you are familiar with. He suggested I might be interested in one or more of your product lines and directed me to your Web site.

從何處得知聯絡資訊

Having looked at your online catalog, I am particularly impressed with your Stack-and-Pack'em storage containers, and I expect they will sell well in all our outlets. My only concern is availability. We would like to begin with a large order (at least 1,500 units) of all sizes in blue, white and red. Do you have enough of those colors in stock? We need them to be delivered no later than March 15.

對產品感興趣

提出顧慮

Since your Web site did not list rates for your products, could you please send me a comprehensive[2] price list? If you have wholesale discounts available for large orders, please include a description

索取資料

上班族加油站

a chain of stores 是指「連鎖店」，亦稱為 chain stores，像超商、飯店等連鎖店就可稱為 supermarket chains、hotel chains。outlet 可稱為「銷售通路」或「暢貨中心」，為專賣過季或折扣商品的商店。

with the price list. Thank you very much. I look forward to hearing back from you.

Sincerely,

Alice Chen
Delta Interiors

Vocabulary & Phrases

1. interior [ɪn`tɪriɚ] *n.* 室內 *adj.* 內部的
2. comprehensive [ˌkɑmprɪ`hɛnsɪv] *adj.* 綜合的

中文翻譯

日期：二〇一二年二月十二日
收件人：尤琳達．菲爾普斯
寄件人：Sadams@deltainteriors.com
主旨：產品詢問：堆疊式收納箱

親愛的菲爾普斯小姐：

我的名字是陳愛麗絲，在台灣擁有一家叫做三角室內裝潢的連鎖店。我們專門解決室內裝潢實際會遇到的問題。我從克里斯多福．荷登那裡取得妳的聯絡資訊，我想妳應該跟他很熟。他認為我應該會對你們的一些產品系列有興趣，於是指引我到你們網站上看看。

看過你們的線上型錄之後，我對你們的堆疊式收納箱印象特別深刻，也預期它們在我們的銷售據點會賣得很好。我唯一擔心的是存貨問題。我們想要先從大筆訂單開始（至少一千五百組）包括藍的、白的、紅的各種尺寸。這些顏色你們有足夠的庫存嗎？這些商品需要在三月十五日前到貨。

由於網站上並未列出商品的價格，妳可以寄給我一份綜合價目表嗎？如果大筆訂單享有批發折扣價，請在價目表附上說明。感激不盡。敬候佳音。

謹上

陳愛麗絲
三角室內裝潢

Ⅳ. 如何用 e-mail 回覆產品詢問

To: Alice Chen

Subject: Re: Product Inquiry: Stack and Pack'em Containers

Dear Ms. Chen,

Thank you for expressing interest in our Stack-and-Pack'em storage containers. I'm happy to inform you that we do have plenty of the colors you requested in stock and available for purchase. However, in order to process your request we will need the address and telephone of your warehouse or distributor. In addition to the price list you asked for, I've also attached a purchase order which you will need to fill out to complete your order with us. Concerning your time constraints,[1] shipping to Taiwan should take no more than a week, so there shouldn't be any worries.

As for wholesale discounts, we are currently not offering any due to higher production costs. But don't let that discourage you. We appreciate your business! Prices fluctuate[2] and regular clients are offered discounted rates on a case-by-case basis.[3]

Regards,

Yolinda Phelps

Creative Plastics

上班族加油站

不要害怕拒絕客戶的要求哦！不過在婉拒的同時別忘了感謝對方，以建立長久的合作關係。

Vocabulary & Phrases

1. constraint [kən`strent] *n.* 約束；限制
2. fluctuate [`flʌktʃə͵wet] *v.* 波動
3. on a case-by-case basis 就個別的情況來看

中文翻譯

收件人：陳愛麗絲
主旨：回覆：產品詢問：堆疊式收納箱

親愛的陳小姐：

感謝妳對我們的堆疊式收納箱感興趣。我很高興通知妳，妳需要的顏色我們還有許多現貨可供銷售。然而，為了處理妳的需求，我們需要妳倉庫或經銷商的地址及電話。除了妳所需要的價目表外，我也附上訂購單，妳必須填寫才算完成訂購。考量到你們的時限，貨運到台灣應該不會超過一個星期，所以應該不用太擔心。

至於批發折扣價，因為生產成本較高所以我們目前並沒有提供。但不要因此感到失望。我們很感謝此次的交易！價格會依照每次的狀況有所變動，並且會有提供老客戶折扣價。

謹上

尤琳達・菲爾普斯
創意塑膠

Try it! 換你試試看!

1. 我從 Judy Jenkins 那裡得到你的聯絡資訊，我想你應該跟她很熟。

2. 我對你們的 Art-Deco 壁貼印象特別深刻。

(壁貼 wall sticker)

3. 我唯一擔心的是產品安全。我們需要它們符合 CPSC 標準。

(符合 conform to)

4. 你可以提供我一份價格資訊嗎？

答案請參閱第 371 頁

V. 用電話詢問產品一定要會這樣說

MP3 TRACK 92

從何處得知聯絡資訊

A **I got your number from** Erica Wheldon, **whom I think you know.**

我想你應該知道艾瑞卡·威爾登，我從她那裡得知你的號碼。

B Yes, we're very good friends.

是的，我們是很好的朋友。

對產品感興趣

A **Which line of products were you interested in specifically?**

你有沒有特別對哪一系列的產品感興趣？

B I'm especially interested in the Roller-Derby Race Cars.

我對 Roller-Derby 賽車系列尤其感興趣。

回答產品問題

A I would like know a little more about your Roller-Derby Race Cars.

我想要多知道一些關於你們 Roller-Derby 賽車系列的情報。

B **Sure thing. I'd be happy to answer any questions you may have.**

沒問題。我很樂意回答你的任何問題。

提出疑慮

A **My only concern is** the delivery date. We need our order to arrive no later than the 21st of June.

我唯一擔心的是交貨日。我們訂的貨必須在六月二十一日之前抵達。

B I think we can handle that.

我想我們可以處理。

索取資料

A **Could you send me** the pricing information for those items?

你可以寄給我那些商品的價格資訊嗎？

B Certainly. I'll get them out to you as soon as possible.

當然。我會儘快寄給你。

正面回應

A **I'm happy to inform you that** our entire stock is on sale.

我很高興通知你，我們所有存貨都在促銷中。

B That's great news. I guess it's time to put in an order.

那真是個好消息。我看要趁現在下單。

VI. 如何用電話詢問產品

三合一廚具的蒂娜致電給廚具供應商代表衛斯里，詢問產品的相關訊息。

LISTENING 請聽 MP3 TRACK 93 ☐ SPEAKING 請跟著 MP3 唸唸看 ☐

Tina: Hello? May I speak with Mr. Wesley Cantor, please?

Wesley: Speaking. Who may I ask is calling?

從何處得知聯絡資訊

Tina: My name is Tina Sanders from Trinity Kitchens. I was given your number by Shelly Thompson. She's a client of yours.

Wesley: That's right. What can I do for you Ms. Sanders?

對產品感興趣

Tina: Well, she tells me that you have the lowest prices on No-Scratch pots and pans in the county, and I was wondering which models you carry.

Wesley: I see. If you like, I could send you some literature on the models we have in that product line.

索取資料

Tina: That would be great. Oh, one more thing before I forget: Can you send me the pricing information for your products as well?

詢問聯絡方式

Wesley: You bet. Why don't you give me your address, and I can get all that out to you this afternoon.

Tina: OK. Shelly gave me your e-mail address as well, so how about I just send it there?

Wesley: Sounds like a plan. I'll keep an eye out for it.

Tina: Great. Thanks a lot.

上班族加油站

sounds like a plan 等同於 sounds good，表示聽起來不錯、聽起來像是一回事。

你好？我可以和衛斯里．康德說話嗎，麻煩你了。

我就是。請問是誰打來的？

我是三合一廚具的蒂娜．山德斯。我從雪莉．湯普森那裡得知你的電話。她是你的客戶。

是的。山德斯小姐，有什麼可以為妳效勞的嗎？

嗯，她跟我說你們有全郡最低價的不留刮痕鍋具組，我想知道你們有哪些款式。

我瞭解了。如果妳想要的話，我可以給妳一些那個產品系列款式的文宣。

那太好了。喔，在我忘記前還有一件事：你可以同時寄給我一份你們的產品價目表嗎？

當然。還是妳給我地址，我今天下午將全部的資料寄給妳。

好的。雪莉也有給我你的電子郵件地址，還是我寄到那裡？

聽起來不錯。我會留意的。

太好了。非常感謝。

Try it! 換你試試看!

WRITING	請依提示寫出完整句子	☐
LISTENING	請聽 MP3 TRACK 94	☐
SPEAKING	請跟著 MP3 唸唸看	☐

1. Ⓐ How did you hear about us?

   Ⓑ ______________________

   ____

   (receive / information / Cathy Fletcher)

   Ⓐ Oh, how nice. Cathy is a long-time client of ours.

2. Ⓐ Which product are you interested in specifically?

   Ⓑ ______________________

   ____

   (be particularly interested in / Noland swivel fans)

3. Ⓐ We have all the colors you requested in stock.

   Ⓑ ______________________

   ____

   (only concern / size / extremely small units)

   Ⓐ Small sizes are available.

4. Ⓐ If you want, I could send you a brochure of our products.

   Ⓑ ______________________

   ____

   (also / pricing information / those products)

答案請參閱第 371 頁

UNIT 17 詢問價格

Price Inquiry

I. 詢問價格一定要會的單字片語

MP3 TRACK 95

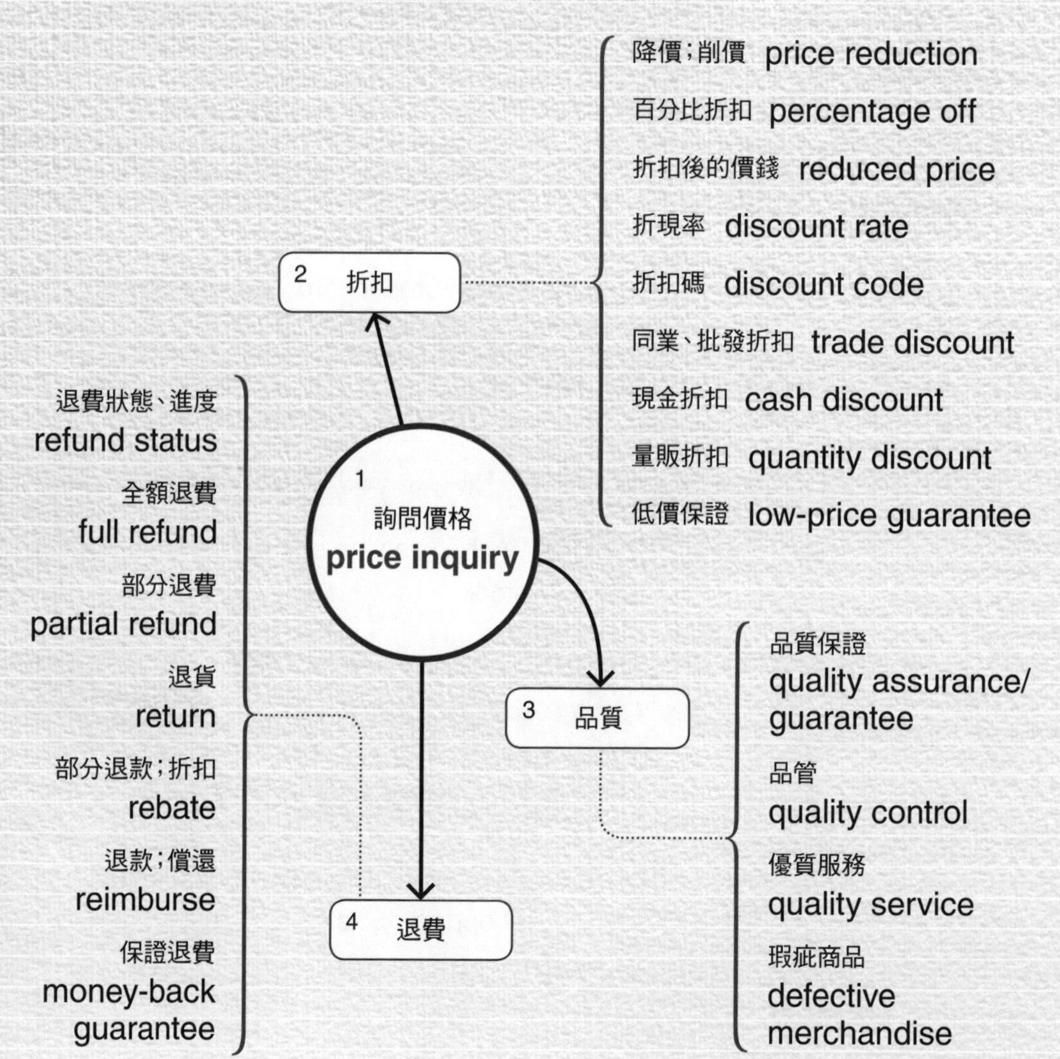

Ⅱ. 詢問價格一定要會的句型

句型 1 考慮下單

We're currently considering placing an order for sth.

例 We're currently considering placing an order for your Super Soft Travel Pillows.
我們目前正在考慮向你們訂購超柔軟旅行用枕。

句型 2 能否提供折扣

Is there any way we could + V.?

例 Is there any way we could negotiate a discount?
有沒有商議折扣的可能呢？

句型 3 提出疑問

We would like to know sth.

例 In particular, we would like to know how defective merchandise will be dealt with.
尤其，我們想知道瑕疵品會怎麼處理。

句型 4 表示願意提供折扣

As far as a discount is concerned, we are willing to negotiate.

例 As far as a discount is concerned, we are willing to negotiate.
說到折扣，我們願意商議。

→ as far as sth is concerned 表示「說到、談到某事」的意思，sth 也可以換成 sb，表示「就某人而言」。

Ⅲ. 如何用 e-mail 來詢價

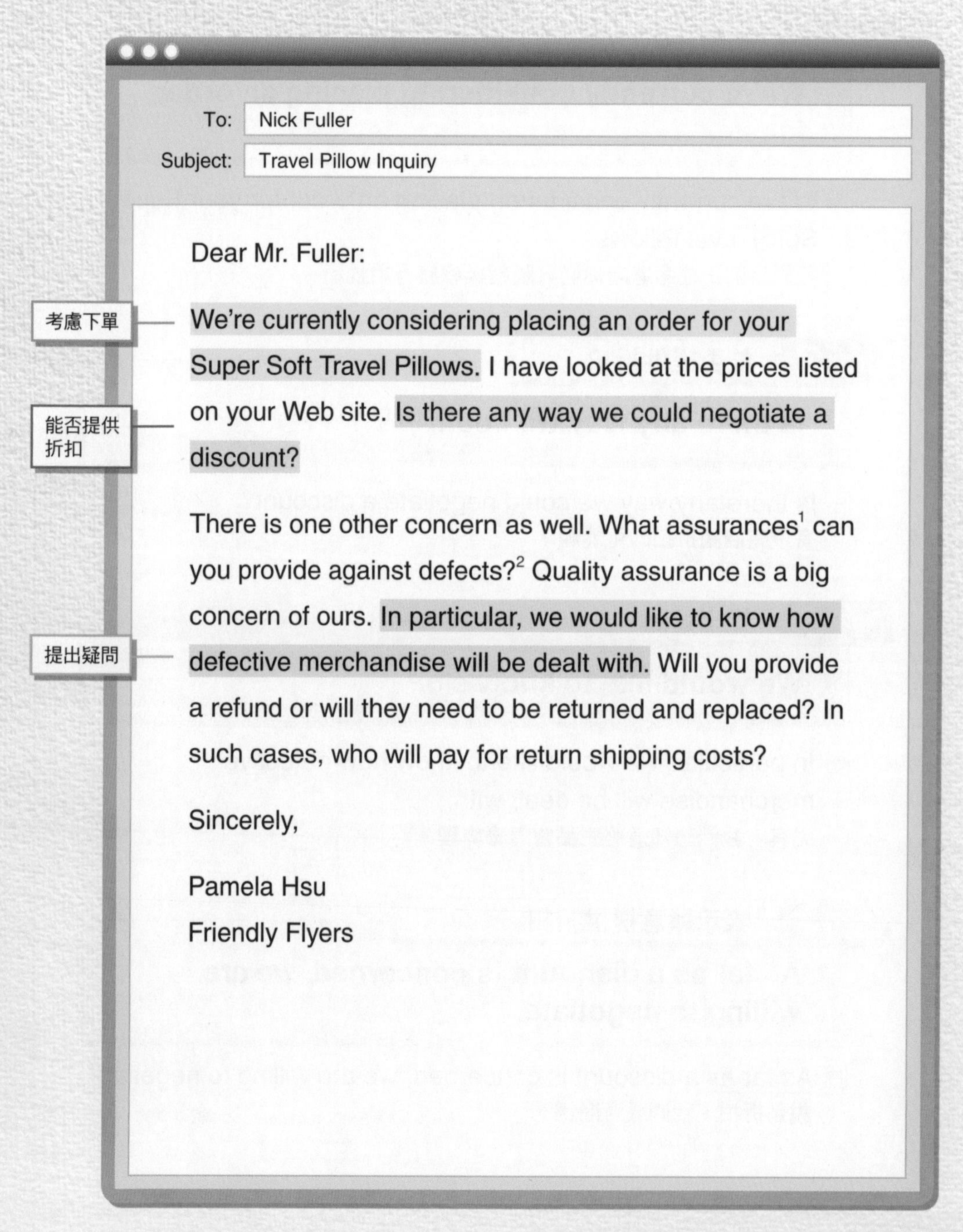

To: Nick Fuller

Subject: Travel Pillow Inquiry

Dear Mr. Fuller:

考慮下單

We're currently considering placing an order for your Super Soft Travel Pillows. I have looked at the prices listed on your Web site. Is there any way we could negotiate a discount?

能否提供折扣

There is one other concern as well. What assurances[1] can you provide against defects?[2] Quality assurance is a big concern of ours. In particular, we would like to know how defective merchandise will be dealt with. Will you provide a refund or will they need to be returned and replaced? In such cases, who will pay for return shipping costs?

提出疑問

Sincerely,

Pamela Hsu

Friendly Flyers

Vocabulary & Phrases

1. assurance [əˋʃurəns] *n.* 保證；擔保
2. defect [ˋdiˏfɛkt] *n.* 瑕疵

中文翻譯

收件人：尼克．富勒
主旨：詢問旅行用枕

親愛的富勒先生：

我們目前正在考慮向你們訂購超柔軟旅行用枕。我已經看過你們網站上所列的價格。有沒有商議折扣的可能呢？

此外還有另一項顧慮。針對瑕疵品，你們能提供怎樣擔保呢？品質保證是我們很重視的問題。尤其，我們想知道瑕疵品會怎麼處理。你們會提供退款，還是商品必須退回更換？如果是這種情形，誰需要負擔退貨的運費呢？

謹上

徐潘蜜拉
友善飛行

文化補給站

詢價技巧

詢價時該如何和對方說我們有預算考量呢？如果你的暗示太過隱晦，對方可能不會注意到，但如果太過直接，可能會讓局面變得很難看。

所以詢問價錢時除了 How much does it cost? 這種制式化的提問，你或許可以技巧性地把問題放在上下文中，不僅能明確傳達出你的預算考量，也不會失禮。

例 Since we've set aside a limited budget, could you please give me a price estimate?
因為我們的預算有限，你可以給我一份估價單嗎？

例 I need to prepare a cost analysis for this project, so could you send me an estimate of how much this is going to cost?
我需要準備這項計畫的成本分析，所以你能將預計的花費寄一份估價單給我嗎？

例 Price will be a critical factor in our decision-making process, so could you please let us know about your discount policy?
在我們的決策過程中，價格是一項很重要的因素，所以可以讓我們知道你們的折扣政策嗎？

Ⅳ. 如何用 e-mail 回覆詢價

To: Pamela Hsu
Subject: Re: Travel Pillow Inquiry

Dear Ms. Hsu,

We are pleased that you are considering making an order for our Super Soft Travel Pillows.

Quality assurance is a concern we share. Rest Easy maintains[1] a defect-free policy. We guarantee complete refunds for defective merchandise. That means there is no need for returns or subsequent[2] worries about shipping costs. Most, if not all, defective merchandise is typically caught by our own QA team. We are confident you and your customers will be satisfied with the finished product.

表示願意提供折扣

As far as a discount is concerned, we are willing to negotiate. How large of an order were you considering? The larger the order, the greater the discount we'd be willing to offer.

Sincerely,

Nick Fuller

Rest Easy

Vocabulary & Phrases

1. maintain [men`ten] *v.* 維持
2. subsequent [`sʌbsɪkwənt] *adj.* 接下來的；繼之而起的

中文翻譯

收件人：徐潘蜜拉
主旨：回覆：詢問旅行用枕

親愛的徐小姐：

我們很高興妳正考慮訂購我們的超柔軟旅行用枕。

品質保證是我們共同關切的事情。高枕無憂一向維持無瑕疵政策。針對瑕疵商品，我們保證全額退費。也就是說，不需要退回商品或擔心後續的運費問題。就算不是全部，大部分的瑕疵品通常都會被我們的品管人員揪出來。我們有信心貴公司及你們的客人對成品會非常滿意。

說到折扣，我們願意商議。你們考慮的訂單量有多大呢？訂單量越多，我們願意提供的折扣就越多。

謹上

尼克· 富勒
高枕無憂

Try it! 換你試試看!

1. 我們正在考慮訂購你們的一些產品。

____

____

____

2. 如果我們的訂單量大的話，可以協議折扣嗎？

____

____

____

3. 對於瑕疵品，你們能給我們什麼保障呢？

____

____

____

4. 你們會如何處理瑕疵品呢？

____

____

____

答案請參閱第 371 頁

V. 電話詢問價格一定要會這樣說

MP3 TRACK 96

有意訂購

A: You've reached Alpine Apparels. This is Cathy speaking. How may I help you?

你所接通的是山岳服飾。我是凱西，有什麼可以為你服務的嗎？

B: Hello Cathy. **We're interested in placing an order for** the Polar Fleece jackets.

哈囉，凱西。我們有興趣向你們訂購極地絨毛夾克。

商量折扣

A: **Is there any way you could lower the price** if we increased our order?

如果我們增加訂單，你們有沒有可能降價呢？

B: That depends. How many units are you willing to commit to?

視情況而定。你們願意承諾多少組呢？

提出疑問

A **What kind of quality assurance standards do you have?**

你們有什麼樣的品質保證標準呢？

B All of our merchandise is inspected and tested before leaving the factory.

我們所有的商品在出廠前，都會經過檢查和測試。

A What do you need to know exactly?

你還需要知道哪些確切的事呢？

B **We need to know precisely** when the order will be delivered.

我們需要知道訂單送達的確切時間。

A **We are curious as to** the exact shipment date.

我們想知道確切的出貨日期。

B Let me look that up for you.

我來幫你看一下。

A Do you have any other questions?

你還有任何其他問題嗎？

B Yes, **we would like to find out** the exact size of the shipping container.

有，我們想要知道貨櫃確切的尺寸大小。

VI. 如何用電話詢問價格

及時鐘錶零件行的麥克打電話給薇拉詢問商品訂購及價錢。

LISTENING 請聽 MP3 TRACK 97 ☐ | SPEAKING 請跟著 MP3 唸唸看 ☐

Vera: Hello. This is Vera Johnson.

Mike: Hi, Vera. This is Mike over at Timely Time Pieces. We're looking at putting in an order for some of your Rise and Shine alarm clocks.

考慮下單

Vera: Oh, the new Rise and Shine model. We have those in stock. How many did you want?

Mike: Well, we were hoping you could cut a little off the top if we placed a large order.

能否提供折扣

Vera: We could do that. Place an order of 5,000 units, and you'll get a 15% discount. How does that sound?

Mike: It sounds great, but I'll still have to talk to my people about it.

Vera: Of course. Was there anything else you needed?

Mike: Actually, yes. We'd like to know how many decibels the alarm sounds at.

提出疑問

Vera: I don't have that information on hand. Can I get back to you on that?

Mike: Naturally. I know it's something of a strange question.

Vera: I'll find out and get back to you soon.

Mike: That's great. I really appreciate it.

哈囉，我是薇拉·強森。

嗨，薇拉。我是及時鐘錶零件行打來的麥克。我們正在考慮訂購你們的起床閃耀鬧鐘。

喔，那款新的起床閃耀鬧鐘。我們目前有存貨。你要多少？

嗯，如果我們下的單量大的話，希望價錢上可以砍一點。

我們做得到。訂購五千件的話打八五折。聽起來如何？

聽起來很棒，但我還是得跟我們其他人談一下。

當然。還需要什麼嗎？

事實上，有的。我們想知道鬧鐘的分貝是多少。

目前我手邊沒有那項資料。我可以之後回覆你嗎？

當然，我知道這個問題有點怪。

我會找出來然後儘快回覆你。

那太好了。真的很感激。

Try it! 換你試試看!

WRITING	請依提示寫出完整句子	☐
LISTENING	請聽 MP3 TRACK 98	☐
SPEAKING	請跟著 MP3 唸唸看	☐

1. Ⓐ How may I help you today?

   Ⓑ ______________________________

   ____

   (interest / place an order / HearWell headphones)

2. Ⓐ How large of an order were you thinking of?

   Ⓑ ______________________________

   ____

   (order / one thousand units / give / discount)

3. Ⓐ Did you have any other questions?

   Ⓑ ______________________________

   ____

   (yes / know / guarantee / defects)

4. Ⓐ You'll receive your order very soon.

   Ⓑ ______________________________

   ____

   (know / precisely / order / be shipped)

答案請參閱第 371 頁

UNIT
18 報價後續追蹤
Price Quote Follow-up

I. 報價後續追蹤一定要會的單字片語 MP3 TRACK 99

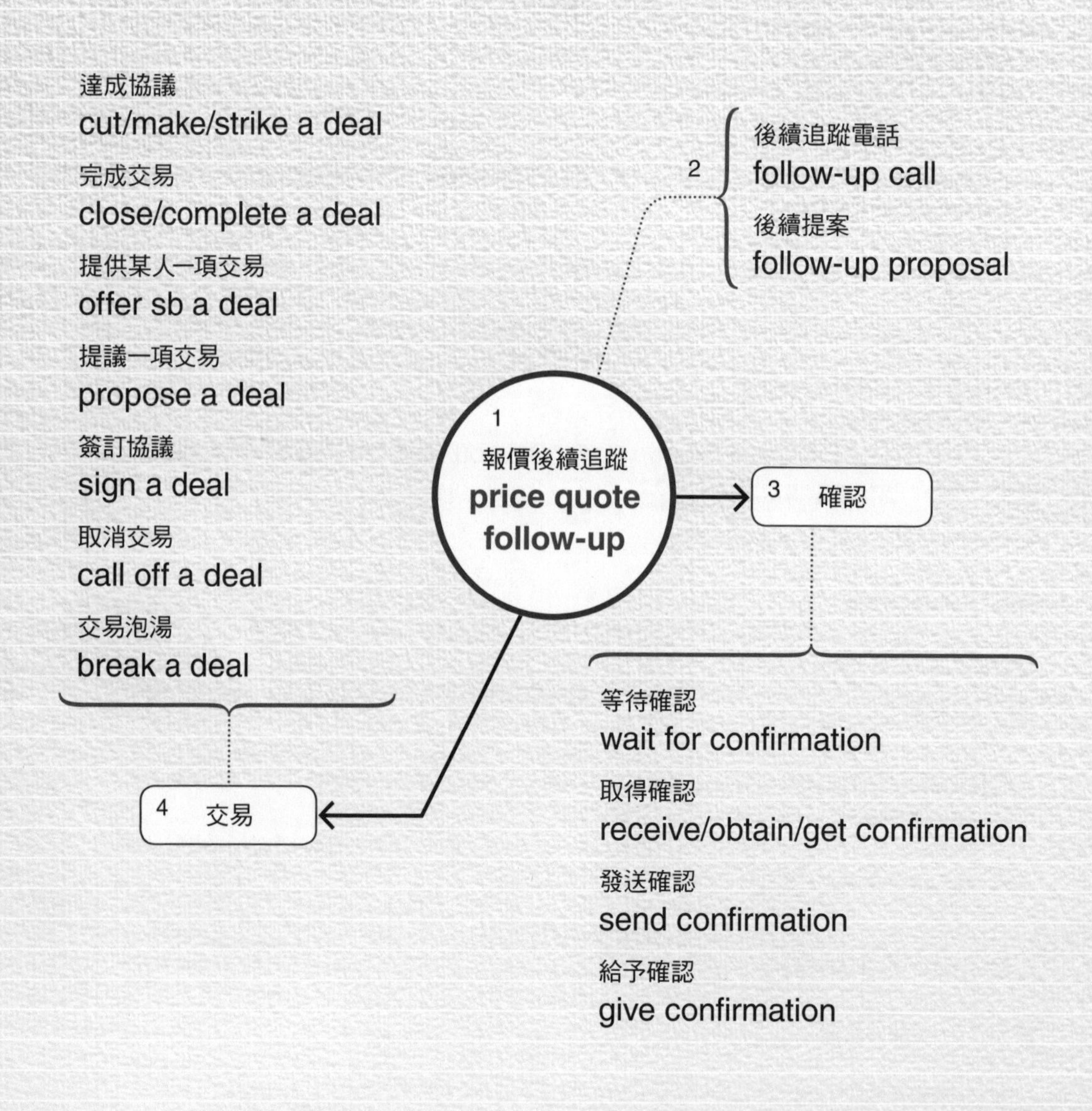

Ⅱ. 報價後續追蹤一定要會的句型

句型 1 等待訂單確認

We are still awaiting your confirmation.

例 We are still awaiting your confirmation.
我們還在等妳的確認。

句型 2 推銷產品

There isn't a more adj. + 商品 + **available.**

例 There isn't a more popular or more comfortable pillow available.
你找不到更受歡迎或更舒適的枕頭了。

句型 3 期待合作

We look forward to hearing back from sb.

例 We look forward to hearing back from you soon.
我們期待儘快聽到妳的回覆。

句型 4 延遲原因

Our offices have been closed for + 原因

例 Our offices have been closed for the extended Chinese New Year holiday.
我們因為中國新年連假而沒有上班。

Ⅲ. 如何用 e-mail 做報價後續追蹤

Date:	February 3, 2012
To:	Pamela Hsu
From:	nfuller@resteasy.com
Subject:	Quotation Follow-up

Hello again, Ms. Hsu,

I'm writing again to follow-up on the previous quotation we sent you for the Travel Pillows you inquired about. I've attached the document to this e-mail, but here is a quick summary of our offer:

Item: Super Soft Travel Pillows **Qty:** 5,000

Total Price: US$6,000.00 **Discount:** 10%

I originally sent this quotation to you on January 17. Perhaps you didn't receive it. We are still awaiting your confirmation. We are very eager to know if you would like to proceed with[1] the order. If the terms are acceptable, please let us know and we will ship it out to you immediately.

等待訂單確認

As you probably know, our Super Soft Travel Pillows are the top-selling travel pillow on the market. There isn't a more popular or more comfortable pillow available. You will have trouble keeping this item on your shelves.

推銷產品

上班族加油站

can't keep this item on your shelves 字面意思是「商品無法在架上留太久」，即表示「熱銷」的意思。

We appreciate your business. If there is anything we can do to facilitate[2] your order, we would be pleased to assist you in any way.

期待合作 — We look forward to hearing back from you soon.

Sincerely,

Nick Fuller

Rest Easy

Vocabulary & Phrases

1. proceed with [pro`sid] *v.* 繼續進行
2. facilitate [fə`sɪlə,tet] *v.* 幫助；促進

中文翻譯

日期：二〇一二年二月三日
收件人：徐潘蜜拉
寄件人：nfuller@resteasy.com
主旨：報價後續追蹤

再次向妳問候，徐小姐：

我再次寫信是要追蹤上次在妳詢問後寄出的旅行用診的報價。我把檔案附在信件中，不過我在這裡快速列出我們報價的摘要：

品項：超柔軟旅行用枕　**數量：**五千　**總價：**六千美元　**折扣：**九折

原本我在一月十七日將這份報價寄給妳。可能妳沒有收到。我們還在等妳的確認。我們很想知道妳是否還想繼續這份訂單。如果這些條件能接受的話，請通知我們，我們將立即出貨給妳。

妳可能知道，我們的超柔軟旅行用枕是市面上最暢銷的旅行用枕。妳找不到更受歡迎或更舒適的枕頭了。這件商品將會熱銷，不會佔架太久的。

我們很感激這筆生意。如果在訂單方面有任何可以幫忙的地方，我們會很樂意提供協助。我們期待儘快聽到妳的回覆。

謹上

尼克· 富勒
高枕無憂

Ⅳ. 如何用 e-mail 回覆報價追蹤信

Date: February 4, 2012
To: Nick Fuller
From: pam_hsu@friendlyflyers.com
Subject: Re: Quotation Follow-up

Dear Mr. Fuller

Please accept my apologies for not getting back to you sooner. Our offices have been closed for the extended[1] Chinese New Year holiday. As such, we did not receive your e-mail in time to reply meaningfully. We are still thinking over your offer. Perhaps we can discuss the details over the phone tomorrow? We would like to seal[2] the deal as soon as possible.

延遲原因

Sincerely,

Pamela Hsu
Friendly Flyers

上班族加油站

利用被動語態 have been closed 可避免將錯誤明確歸咎在某人或某事上，表示較客觀的原因。

Vocabulary & Phrases

1. extended [ɪk`stɛndəd] *adj.*
延長的；擴張的

2. seal [sil] *v.*
確認；保證（契約、交易）

中文翻譯

日期：二〇一二年二月四日
收件人：尼克・富勒
寄件人：pam_hsu@friendlyflyers.com
主旨：回覆：報價後續追蹤

富勒先生你好：

很抱歉沒有儘快回信給你，請接受我的道歉。我們因為中國新年連假而沒有上班。因此，我們沒有及時收到你的來信並做有意義的回覆。我們還在考慮你的開價。或許我們可以明天在電話上談？我們想要儘早達成協議。

謹上

徐潘蜜拉
友善飛行

Try it! 換你試試看!

1. 我們仍等候聽到你對此訂單的最後決定。

2. 你們在別處找不到更好或更值得信賴的電池了。

3. 我們希望可以在下星期之前收到你們的回覆。

4. 我們辦公室因為颱風而沒有上班。

答案請參閱第 372 頁

V. 電話報價後續追蹤一定要會這樣說 MP3 TRACK 100

等待訂單確認

A **We were wondering if you still wanted to go ahead with the order.**

我們想知道這份訂單是不是還要繼續處理呢？

B We do. Sorry about the delay.

要的。很抱歉耽擱了。

A I'm sorry I couldn't/didn't call you back sooner.

抱歉我沒有儘快回電給你。

B That's OK. **We're just waiting for your confirmation.**

沒關係。我們只是在等你確認。

推銷產品

A **I don't know if you were having any second thoughts, but you won't find a better model on the market.**

我不知道是否你有其他想法，不過你在市面上找不到更好的款式了。

B That's true. It is the best product out there. / Well, I'm not so sure about that.

沒錯，這的確是最好的產品。／嗯，對此我不是很確定。

對原本決定的事情有了其他想法可用 have second thoughts about sth 來表示。

給予協助

A **We'll be happy to help you with any questions you might have about your order.**

你在訂單上有任何疑問我們都很樂意為你解決。

B Actually, I do have a few things I'd like to ask you about.

事實上，我的確有一些事情想要問你。

延遲原因

A **Our Internet service has been down for the past few days.** We haven't been able to connect to the server.

我們的網路服務在過去幾天一直出問題。我們一直無法連上伺服器。

B That's too bad. No wonder we haven't heard back from you. Have you got it sorted out yet?

那太糟了。難怪我們沒有收到你的回覆。你們解決了嗎？

sort out 為口語用法，表示「解決；弄清楚」，在此也可以用 fix 或 work out 來代替。

A We haven't heard back from you in a while.

我們有一陣子沒收到您的消息了。

B I'm really sorry about that. **We've been really busy here lately.**

我真的很抱歉。我們這邊最近真的很忙。

VI. 如何用電話做報價後續追蹤

帽子供應商查克打電話給泰咪做報價後追蹤，並再次強調產品優勢。

LISTENING 請聽 MP3 TRACK 101 ☐	SPEAKING 請跟著 MP3 唸唸看 ☐

Tammy: You've reached the Hat House. This is Tammy Tanner. How can I help you?

Chuck: Hi, Tammy. It's Chuck from Henderson's Hats.

Tammy: Hello, Chuck. I've been meaning to get back to you. We've had a bit of a busy week. Sorry about that.

延遲原因

Chuck: It's OK. I know you said you needed that order we talked about, but we haven't gotten a confirmation yet. Should I ship it out?

等待訂單確認

Tammy: Here's the thing. We're having second thoughts about the new hats. We're not sure they're going to sell well.

Chuck: You should give them a chance, Tammy. There isn't a more comfortable winter hat out there.

推銷產品

Tammy: I'm sure they're comfortable. What concerns me is how many I can sell.

Chuck: I'm sure we can work something out. Maybe we can send you fewer than the original order.

期待合作

Tammy: That might work. I'll have to think about it.

Chuck: No problem. Why don't you . . .

Tammy: Sorry to interrupt you, Chuck. I've got another call coming in.

Chuck: That's fine. We can finish this later. Bye, now.

Tammy: Thanks a lot. Bye, Chuck.

您撥的是帽子之家。我是泰咪．唐納。有什麼能幫忙的嗎？

嗨，泰咪。我是漢德森帽子的查克。

你好，查克。我一直想回電給你。我們這週有點忙。很抱歉。

沒關係。我知道妳說妳要那份之前談過的訂單，但我們還沒得到任何確認。我該出貨嗎？

重點在這裡。我們對於這些新帽子有不同的看法。我們不確定它們會不會賣得好。

妳該試試看，泰咪。冬天的帽子妳在外面找不到更舒適的了。

我確定它們很舒適。我擔心的是我可以賣掉多少。

我相信我們有解決辦法。或許我們可以寄給妳比原始訂單少一點的量。

那或許可行。我需要考慮一下。

沒問題。妳何不……

查克抱歉打斷你。我有另一通電話進來。

沒關係。去接電話吧。我們可以晚一點再談。先說再見。

非常謝謝你。再見，查克。

Try it! 換你試試看!

WRITING	請依提示寫出完整句子	☐
LISTENING	請聽 MP3 TRACK 102	☐
SPEAKING	請跟著 MP3 唸唸看	☐

1. Ⓐ Thank you for your follow-up call.

   Ⓑ We'd like to know if you'll go ahead with the order.

   ____

   ____

   (not receive / confirmation / you)

2. Ⓐ How is your brand compared to other tissue brands?

   Ⓑ ______________________________

   ____

   (not find / softer / tissue / market)

3. Ⓐ We have a few questions about the order.

   Ⓑ ______________________________

   ____

   (happy / assist / possible)

4. Ⓐ I haven't heard back from you about the items in stock.

   Ⓑ I'm sorry. ______________________________

   ____

   (inventory manager / out sick / past three days)

答案請參閱第 372 頁

UNIT 19 | 議價

Negotiating Prices

I. 議價一定要會的單字片語

MP3 TRACK 103

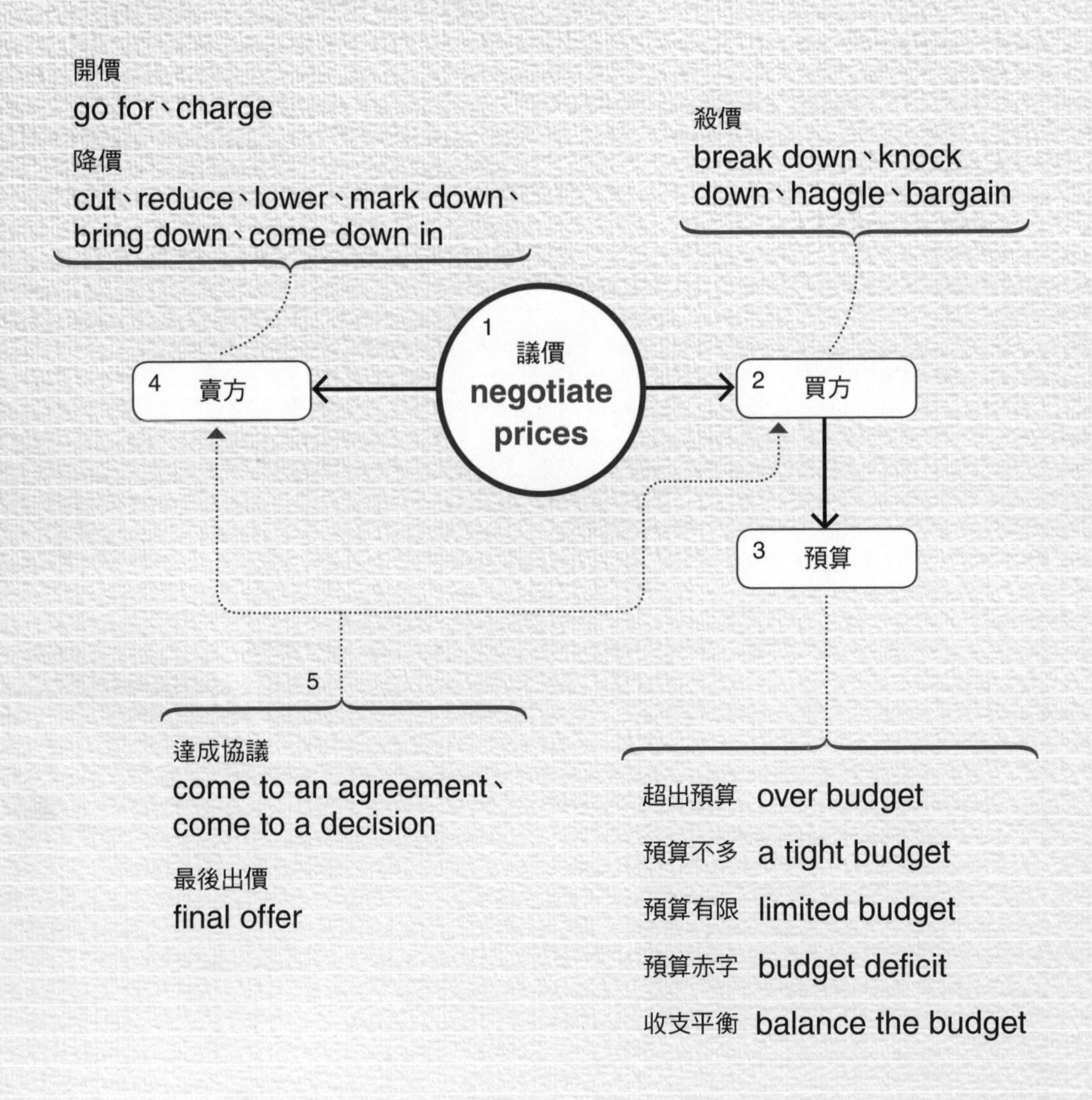

II. 議價一定要會的句型

句型 1 表示超出預算

Unfortunately, sth **does not allow us to meet the quoted price.**

例 Unfortunately, our budget does not allow us to meet the quoted price.
很可惜，我們的預算不容許我們接受這份報價。

句型 2 狀況說明

At the current rate, **we would have to** + V.

例 At the current rate, we would have to lower our order from 4,000 to 3,000 units.
以目前匯率來看，我們必須將訂單從四千降到三千組。

句型 3 議價

Is there any way we could + V.?

例 Is there any way we could get 4,000 units for $60,000?
我們有辦法用六萬元買四千組嗎？

句型 4 期待達成交易

We would like to come to a deal . . .

例 We would like to come to a deal as soon as possible.
我們希望能儘早達成交易。

Ⅲ. 如何用 e-mail 議價

Date: August 30, 2011

To: Paul Rogers

From: Wendy Lin

Subject: Over Our Budget

Greetings Mr. Rogers,

Thank you for your prompt reply. I wasn't expecting to hear back from you so quickly. Unfortunately, our budget does not allow us to meet the quoted price. Due to other investments, our funds are limited. At the current rate, we would have to lower our order from 4,000 to 3,000 units. Since this number is a bit too small for our needs, we are forced to ask for a discount. Is there any way we could get 4,000 units for $60,000?

超出預算

狀況說明

議價

In our past dealings you have known us to be very straightforward.[1] We also appreciate your willingness to accommodate[2] our needs. That is why we come back to do business with your company.

上班族加油站

議價時可表示與對方一直以來都合作愉快，並希望能再次合作，作為議價談判的籌碼。

We are in somewhat of a hurry to receive the units, so we would like to come to a deal as soon as possible. Thank you again.

期待達成交易

Sincerely,

Wendy Lin, Area Manager

Dragon Distributors

Vocabulary & Phrases

1. straightforward [͵stretˋfɔrwɚd] *adj.* 直接坦率的
2. accommodate [əˋkɑmə͵det] *v.* 配合；舒適

中文翻譯

日期：二〇一一年八月三十日
收件人：保羅．羅傑斯
寄件人：林溫蒂
主旨：超出預算

羅傑斯先生你好：

感謝你迅速回覆。我沒有想到你這麼快就回覆了。很可惜，我們的預算不容許我們接受這份報價。因為還有其他投資，我們的資金有限。以目前匯率來看，我們必須將訂單從四千降到三千組。由於這個數量和我們的需求比起來又太少，因此我們被迫要求折扣。我們有辦法用六萬元買四千組嗎？

從我們過去的交易，你必然知道我們是非常直率的。我們也很感激你們願意配合我們的需求。那也是為什麼我們一再和你們做生意的原因。

我們有點急需收到這些商品，所以希望能儘早達成交易。再次感謝你。

謹上

區經理　林溫蒂
龍牌經銷商

文化補給站

每個文化議價的方法都不同。對美國人來說，能夠迅速、實際且果斷達到目的的人才是好的議價人才。相較於其他某些文化，美國的談判節奏是相當快的，他們認為「時間就是金錢」，因此有些買賣甚至在談判的初步階段就得出結果了。

此外，他們認為談判語言應該直接、清楚，事實上，談判過程的回答若是含糊不清可能會被視為不夠真誠。

最後，美國的談判人傾向於將重點放在「回報」上，也就是說，如果能替公司帶來利潤就是好的提議，甚至於能夠在短時間內創造利潤則更好，因此他們不像許多亞洲商人，以建立長期的夥伴關係或創造長期的利潤為主要考量。他們著眼的是短期利潤。將來和美國人談判時，最好留意這一點文化上的差異。

Ⅳ. 如何用 e-mail 回覆議價

Date: August 30, 2011
To: Wendy Lin
From: progers@digideals.net
Subject: Re: Over Our Budget

Dear Ms. Lin,

We understand your budget[1] and time constraints. For the number you mentioned, the best we can offer is 3,500 units. Your business is important to us, so we are willing to offer a onetime[2] discount. If you really need 4,000 units, then our price will be $68,000. Is either of these terms acceptable?

你也可以多學一些好用句喔！

Sincerely,

Paul Rogers, Sales Associate
Digital Deals

Vocabulary & Phrases

1. budget [ˋbʌdʒət] *n.* 預算
2. onetime [ˋwʌnˋtaɪm] *adj.* 一次性的

中文翻譯

日期：二〇一一年八月三十日
收件人：林溫蒂
寄件人：progers@digideals.net
主旨：回覆：超出預算

親愛的林小姐：

我們能理解妳的預算及時間限制。妳所提到的數字，我們最多只能提供三千五百組。你們的生意對我們來說很重要，所以我們願意提供一次性折扣。如果你們真的需要四千組，那我們的價錢會是六萬八千元。這兩個條件哪個可以接受呢？

謹上

銷售員　保羅．羅傑斯
數位買賣

Try it! 換你試試看!

1. 很遺憾地，我們這次訂單的資金有限。

2. 我們必須將此次訂單降到一萬五千組。

3. 有可能提供折扣嗎？

4. 我們希望能快點達成交易。

(達成交易 come to a deal)

答案請參閱第 372 頁

V. 電話議價一定要會這樣說

MP3 TRACK 104

超出預算

A **The price you quoted is a little out of our budget.**
你報的價有點超出我們的預算。

B What did you have in mind?
你預期的價格是多少？

解釋情況

A **Without a discount, we will be forced to lower our order.**
沒有折扣的話，我們得被迫刪減訂量。

B I'm sorry to hear that, but I'm afraid we can't go any lower on the price.
我很遺憾聽到你這麼說，但恐怕我們價格無法再降了。

議價

A **Would it be possible to order** 3,000 units **at** 500,000 NTD?
可不可能以新台幣五十萬訂購三千組呢？

B At that price, the best thing we can offer is 2,500 units.
以那個價錢，我們最多能提供兩千五百組。

期待折扣

A **I can't tell you how much we would appreciate a discount.**

有折扣的話我們將會無比的感激。

B Well, how big of a discount were you thinking of?

嗯，你們想要多少折扣？

拒絕折扣

A Could you lower the price a bit?

你可以把價錢再降一點嗎？

B Unfortunately not. **That is our final offer.**

很遺憾不行。那是我們的底價了。

表達交易意願

A **We are very eager to come to a deal as soon as possible.**

我們非常希望能儘早達成交易。

B I understand. I'll get back to you within the hour. How does that sound?

我瞭解。我會在一小時內回覆你。你覺得如何？

VI. 如何用電話議價

供應商維斯打電話詢問顧客荷璞對開價是否滿意，荷璞表示超出預算並希望能給個折扣。

LISTENING 請聽 MP3 TRACK 105 ☐ | SPEAKING 請跟著 MP3 唸唸看 ☐

Wes: Hello, Hope. Wes here.

Hope: Hi, Wes. I'm glad you called. What did you think of our offer?

超出預算

Wes: Actually, it's a bit higher than we were hoping. Frankly, it's just out of our budget.

Hope: That's too bad. How much were you thinking of spending on the order?

議價

Wes: We're prepared to pay as much as $150,000, but that's really the highest we can go. Could you offer us a discount?

Hope: Hmm. I'll have to get back to you on this. I need to do some calculations.

期待達成交易

Wes: Alright. When do you think you could give us an answer? We're very eager to come to an agreement as soon as possible.

Hope: Would tomorrow morning be soon enough?

Wes: That would be fine.

Hope: OK, then. I'll have a chat with my people, and give you a call first thing tomorrow.

Wes: Thanks a bunch, Hope.

妳好，荷璞。我是維斯。

嗨，維斯。很高興你打來。你覺得我們開的價如何？

事實上，比我們希望的高了點。坦白說，超出我們的預算了。

那太糟了。你們預計要花多少錢在這筆訂單上？

我們打算最多十五萬元，但那是我們能出的最高價了。妳可以提供我們折扣嗎？

嗯。我必須回頭再答覆你這件事。我必須要算一下。

好的。妳什麼時候可以給我們答覆呢？我們非常希望能儘早達成協議。

明天早上夠快了嗎？

那應該可以。

好的，那就這樣。我們內部會先談一下，然後明天第一時間打給你。

非常感謝，荷璞。

Try it! 換你試試看!

WRITING	請依提示寫出完整句子	☐
LISTENING	請聽 MP3 TRACK 106	☐
SPEAKING	請跟著 MP3 唸唸看	☐

1. Ⓐ What did you think of our offer?

   Ⓑ ______________________________

   ____

   (quoted price / little / budget)

2. Ⓐ Did you review the price list we sent you?

   Ⓑ Yes. ______________________________

   ____

   (without / rate adjustment / lower / order)

3. Ⓐ How much were you thinking of spending?

   Ⓑ ______________________________

   ____

   (is / any way / lower / price)

4. Ⓐ May I get back to you in a few days?

   Ⓑ Please let us know at the earliest. ______

   ____

   ____

   (would like / come to a deal / as soon as possible)

答案請參閱第 372 頁

UNIT 20 索取樣品

Request for Samples

I. 索取樣品一定要會的單字片語

MP3 TRACK 107

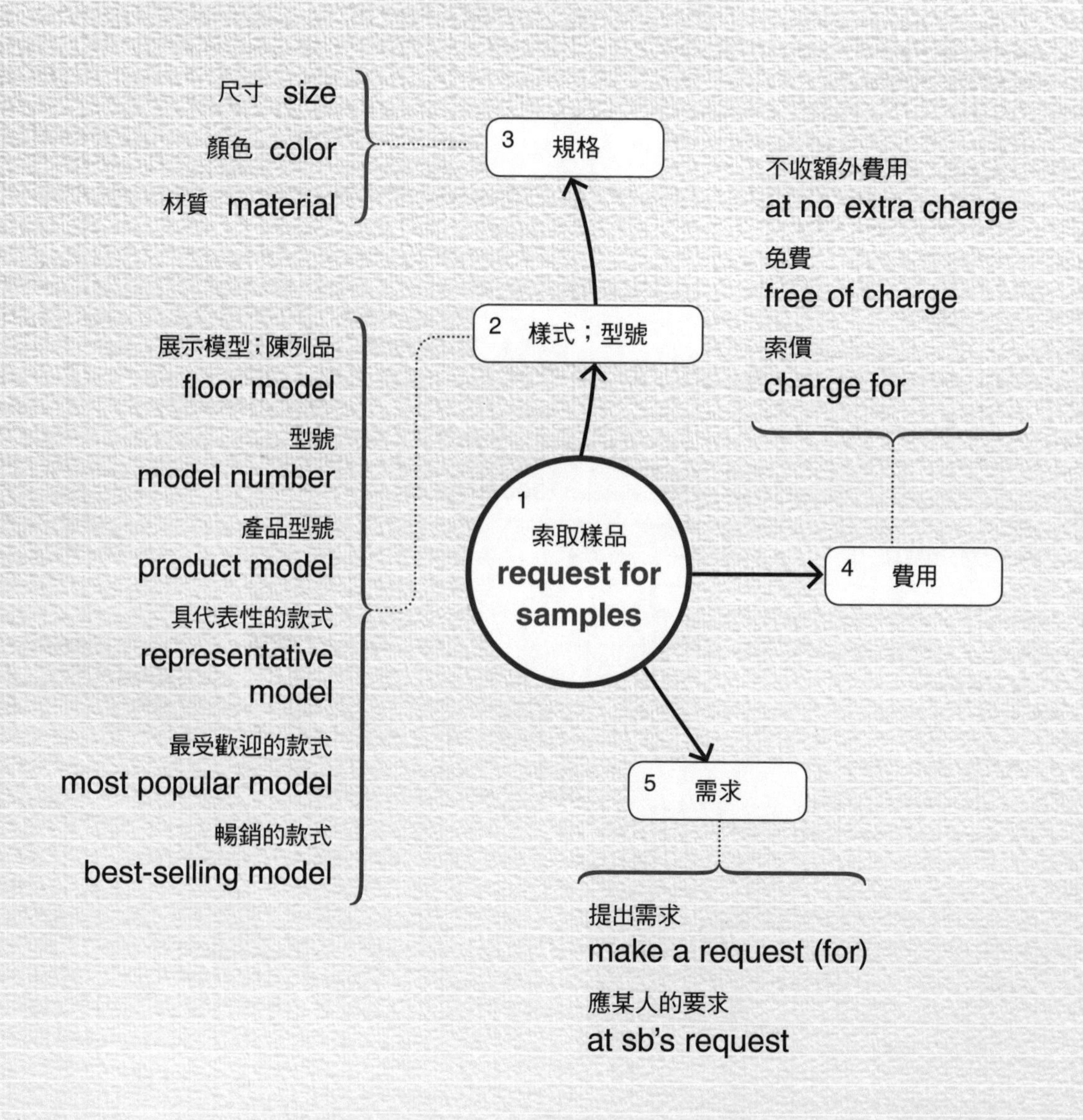

Ⅱ. 索取樣品一定要會的句型

句型 1 表明來意

The intent of this e-mail is + to V.

例 The intent of this e-mail is to request samples of your porcelain floor tiles.
這封信的目的是想向你索取地板磁磚的樣品。

句型 2 樣品需求

Please limit the samples to sth（尺寸、規格）

例 Please limit the samples to those sized 4x4 inches wide or smaller.
請將這些樣品規格限制在四吋見方的大小，或是更小一點。

句型 3 寄送地址

S. **may be delivered to the following address:**

例 The samples may be delivered to the following address:
樣品可以寄到下面地址：

句型 4 詢問費用

Would you like to be reimbursed for the cost of sth?

例 Would you like to be reimbursed for the cost of the samples?
你需要我們支付樣品的費用嗎？

Ⅲ. 如何用 e-mail 索取樣品

Date: October 3, 2011
To: info@ttiles.com.tw
From: ajackson@homestyle.com
Subject: Sample Request

Greetings,

表明來意 — The intent of this e-mail is to request samples of your porcelain floor tiles. We've heard good things about them, but would like to see for ourselves. 樣品需求 — Please limit the samples to those sized 4x4 inches wide or smaller. We're not interested in large tiles. If we like what we see, we plan on placing a sizeable[1] order.

寄送地點 — The samples may be delivered to the following address:

523 Templeton Avenue
Greenwood, MI
07653 USA

詢問費用 — Would you like to be reimbursed[2] for the cost of the samples? We are more than willing to do so. If you prefer to have the samples returned to you, we can do that as well.

Thank you for your time.

Sincerely,

Amanda Jackson, General Manager

Home Style Interiors

Vocabulary & Phrases

1. sizeable [ˋsaɪzəbəl] *adj.* 相當大的；可觀的
2. reimburse [͵riəmˋbɝs] *v.* 償還；歸還

中文翻譯

日期：二〇一一年十月三日
收件人：info@ttiles.com.tw
寄件人：ajackson@homestyle.com
主旨：索取樣品

你好：

這封信的目的是想向你索取地板磁磚的樣品。我們已經聽說許多關於此磁磚的優點，但還是想要親眼瞧一瞧。請將這些樣品規格限制在四吋見方的大小，或是更小一點。我們對大片磁磚不是很有興趣。如果我們對看到的商品滿意的話，我們打算要下一筆可觀的訂單。

樣品可以寄到下面地址：
07653 美國密西根州葛林伍德市
坦普雷頓大道五百二十三號

你需要我們支付樣品的費用嗎？我們非常樂意支付。如果你比較想要我們歸還樣品，我們也可以這麼做。

感謝你撥冗。

謹上

總經理　亞曼達·傑克森
室內居家風格設計

Ⅳ. 如何用 e-mail 回覆樣品索取信

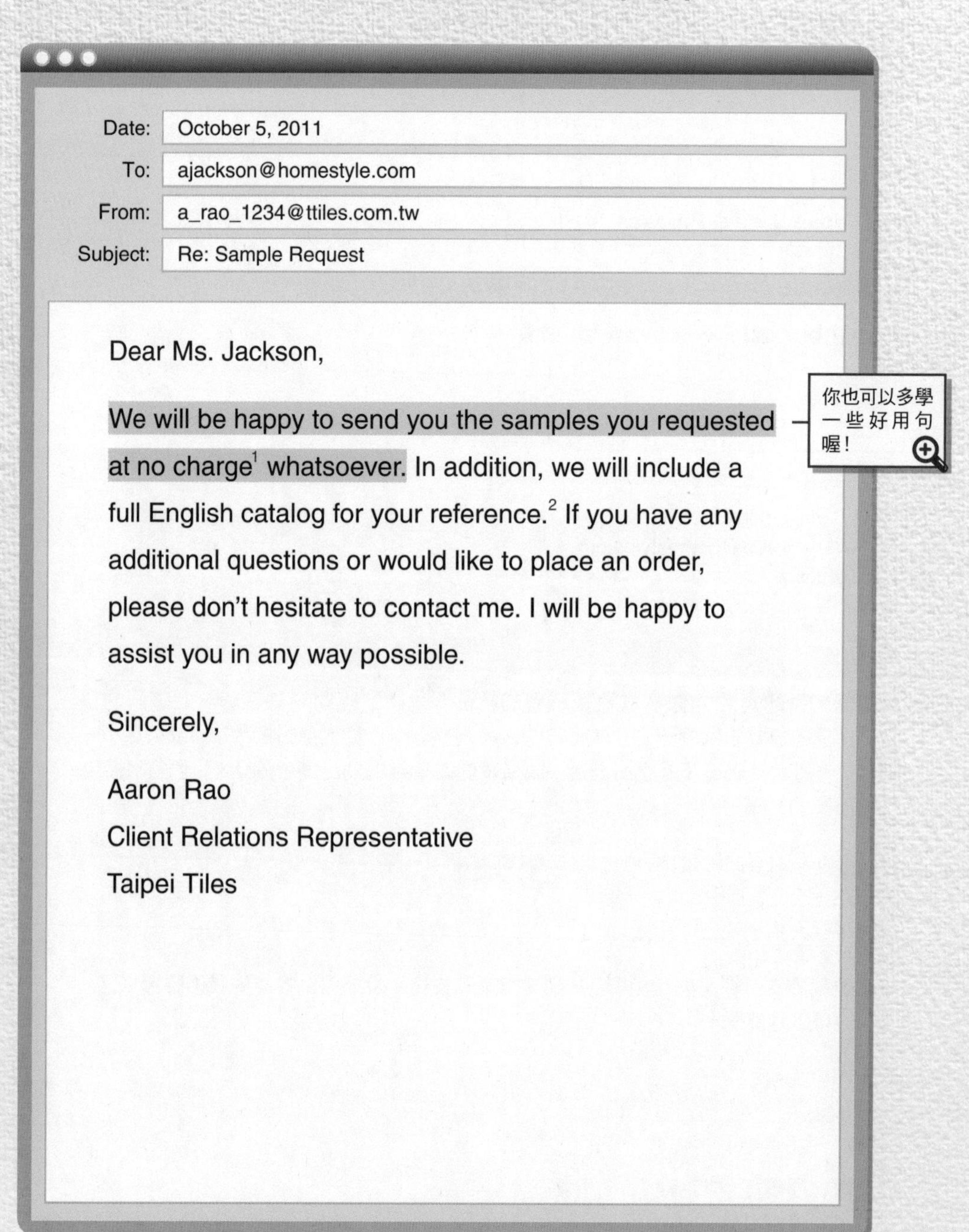
Date: October 5, 2011

To: ajackson@homestyle.com

From: a_rao_1234@ttiles.com.tw

Subject: Re: Sample Request

Dear Ms. Jackson,

We will be happy to send you the samples you requested at no charge[1] whatsoever. In addition, we will include a full English catalog for your reference.[2] If you have any additional questions or would like to place an order, please don't hesitate to contact me. I will be happy to assist you in any way possible.

Sincerely,

Aaron Rao

Client Relations Representative

Taipei Tiles

Vocabulary & Phrases

1. charge [tʃɑrdʒ] *n.* 費用
2. reference [ˋrɛfərəns] *n.* 參考

中文翻譯

日期：二〇一一年十月五日
收件人：ajackson@homestyle.com
寄件人：a_rao_1234@ttiles.com.tw
主旨：回覆：索取樣品

親愛的傑克森小姐：

我們很樂意寄給妳所要求的樣品且不收取任何費用。除此之外，我們附上全英文的型錄供妳參考。如果妳有其他疑問或是想下訂單的話，請不要客氣，儘管與我聯絡。我會很樂意在各方面協助妳。

謹上

饒亞倫
客戶關係代表
台北磁磚

Try it! 換你試試看!

1. 這封信是要向你們索取藍色油漆的樣品。

2. 我們只對金屬款式有興趣。

3. 請將樣品寄到這個地址：台北市忠孝東路六段二〇一號。

(忠孝東路 Zhongxiao East Road)

4. 你需要償還那些樣品的費用嗎？

答案請參閱第 372 頁

V. 用電話索取樣品一定要會這樣說

MP3 TRACK 108

索取樣品

A **The reason I'm calling is to request some samples of** your new dish racks.
我打電話的原因是想索取一些你們新出的碗盤架的樣品。

B Which ones exactly? We've put a few new models out recently.
精確來說是哪一種呢？我們最近推出了幾款新的款式。

A How can I direct your call?
請問要轉哪個部門呢？

B **I'm calling to ask for some samples.**
我打來是想要索取一些樣品。

樣品需求

A **Can you limit the samples to** discount models only?
你可以將樣品只限制在那些折扣商品嗎？

B Of course. We'll get those out to you sometime this afternoon.
當然可以。今天下午我們就會將那些寄去給你。

A Which of our models were you interested in?
你對我們哪些款式有興趣？

B	**We're interested in your line of** protective headgear.	我對你們的防護頭盔系列有興趣。

寄送樣品

A	**Where should we send the samples?** We don't have your address on file.	我們應該將樣品寄到哪裡呢？我們檔案裡沒有你們的地址。
B	Send them to this address. Are you ready?	把它們寄到下列地址。你準備好了嗎？
A	Hold on. Let me grab a pen.	稍等一下。我拿支筆。

詢問費用

A	**Are the samples complimentary or will you be billing us for them?**	這些樣品是免費的還是你會跟我們索取費用？
B	We can send you as many as ten free samples.	我們最多可寄十件免費樣品給你。

這裡的 as many as 也可以用 up to 來表示「最多」。

VI. 如何用電話索取樣品

經銷商代表黛伯拉打電話給布料供應商麥特，希望能索取一些樣品做參考。

LISTENING 請聽 MP3 TRACK 109 ☐ | SPEAKING 請跟著 MP3 唸唸看 ☐

Matt: Good morning. Friendly Fabrics. This is Matt Tyler. How may I help you?

Debra: Good morning. My name is Debra Hsu from Asiatic Distributors. I'm calling to request samples of your fabrics. We'd be interested in making large orders if we like what we see.

表明來意

Matt: We'd be happy to send you some samples. We have a wide range of fabrics available. What sort of material are you interested in?

Debra: A representative sampling would be best. And one more thing: Will there be a charge for the samples?

樣品需求

寄送地點

Matt: We don't charge for samples. They're rather small. Why don't you give me the address, and I'll have them shipped out this week?

詢問費用

Debra: Sounds good. Can I e-mail it to you?

Matt: Sure that would probably be easier. Do you have my e-mail address?

Debra: Let me just verify that I have the right one: matt-underscore-tyler-at-friendly-fabrics-dot-com. Is that right?

Matt: That's me. I'll send you a reply as soon as I get the e-mail.

Debra: That's great, Mr. Tyler. Thanks for your help. I'll let you go now.

Matt: Don't mention it. Have a good day.

早安。友善布料。我是麥特·泰勒。我可以怎麼幫妳？

早安。我是亞洲經銷商的徐黛伯拉。我打來是想索取布料樣品。如果我們看到樣品喜歡的話，就會有興趣下大筆訂單。

我們很樂意寄給妳一些樣品。我們有多種布料可供選擇。妳對哪一種材質有興趣？

具代表性的樣品最好。還有件事：這些樣品有收取費用嗎？

我們樣品是不收費的。它們都很小。妳何不給我地址，然後我在這星期出貨？

聽起來不錯。可以用電子郵件寄給你嗎？

確實那樣或許會更方便。妳有我的電子郵件地址嗎？

讓我確認一下這地址是否正確：matt 底線 tyler-@-friendly-fabrics-dot-com。對嗎？

是我的沒錯。我收到郵件後會盡快向妳回覆。

太好了，泰勒先生。感謝你的幫忙。現在不打擾你了。

不用客氣。祝妳有個愉快的一天。

Try it! 換你試試看!

WRITING 請依提示寫出完整句子	☐
LISTENING 請聽 MP3 TRACK 110	☐
SPEAKING 請跟著 MP3 唸唸看	☐

1. Ⓐ Hello, Mr. Cai. What can I do for you today?

   Ⓑ ______________________________

   ____

   (call / request / samples / novelty clocks)

2. Ⓐ Were you interested in all the different sizes we offer?

   Ⓑ ______________________________

   ____

   (not interested / miniature models)

3. Ⓐ Where should we send the samples? We don't have your address on file.

   Ⓑ ______________________________

   ____

   (send / this address) (do / need / get / pen)

4. Ⓐ There will be a small fee for shipping you the samples.

   Ⓑ That's fine. ______________________________

   ____

   (how / you / us / pay / samples)

答案請參閱第 372 頁

UNIT 21 通知價格調漲
Price Increase Announcement

I. 通知價格調漲一定要會的單字片語

MP3 TRACK 111

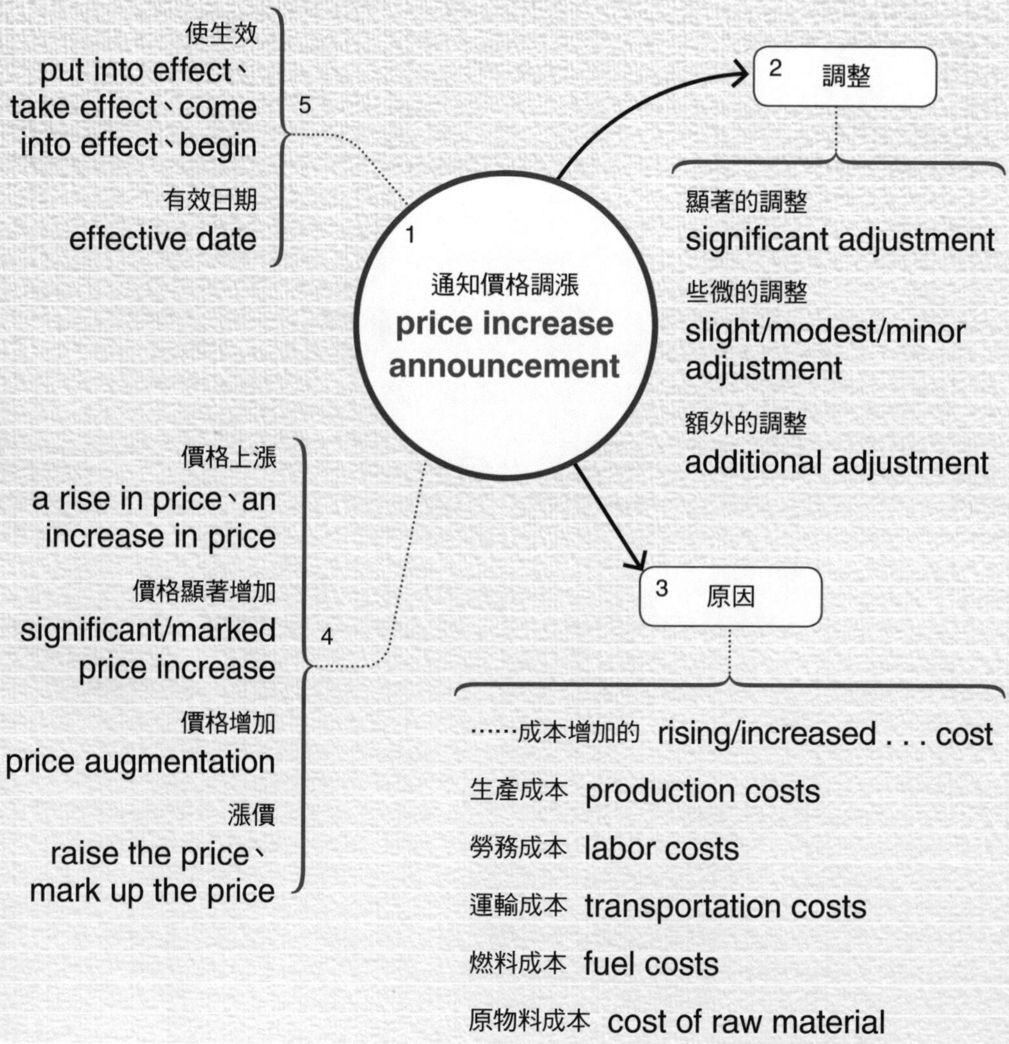

1 通知價格調漲 **price increase announcement**

2 調整
- 顯著的調整 significant adjustment
- 些微的調整 slight/modest/minor adjustment
- 額外的調整 additional adjustment

3 原因
- ……成本增加的 rising/increased . . . cost
- 生產成本 production costs
- 勞務成本 labor costs
- 運輸成本 transportation costs
- 燃料成本 fuel costs
- 原物料成本 cost of raw material

4
- 價格上漲 a rise in price、an increase in price
- 價格顯著增加 significant/marked price increase
- 價格增加 price augmentation
- 漲價 raise the price、mark up the price

5
- 使生效 put into effect、take effect、come into effect、begin
- 有效日期 effective date

II. 通知價格調漲一定要會的句型

句型 1 發出漲價通知

We regret to inform you that . . .

例 We regret to inform you that due to rising costs, we are no longer able to provide our fresh spring water at the same low prices.
我們很遺憾要通知你，因為成本上漲的關係，我們無法再以同樣的低價來提供新鮮泉水。

句型 2 調漲原因

S. + **necessitate a rate adjustment effective** + 日期

例 The growing price of gasoline, combined with increased labor fees, necessitates a rate adjustment effective March 15.
日漸上漲的油價，加上勞務費用的增加使我們必須從三月十五日起做價格調整。

句型 3 強調產品優勢

As you know, S. **is your best source of . . .**

例 As you know, Crystal Springs is your best source of crystal-clean drinking water.
如你所知，水晶泉水是純淨飲用水的最佳來源。

句型 4 招攬訂單

Why not take this opportunity + to V.?

例 Why not take this opportunity to put an order in now while prices are still low?
何不趁價格還低的時候把握機會下訂單呢？

Ⅲ. 如何用 e-mail 通知價格調漲

Date: January 11, 2012
To: Vending Unlimited
From: m_douglas@crystalsprings.com
Subject: Rate Adjustment

Dear Valued Customer,

發出通知

We regret to inform you that due to rising costs, we are no longer able to provide our fresh spring water at the same low prices. In today's economy, it has become impossible to produce and ship such high-quality water without increased costs. The growing price of gasoline, combined with increased labor fees, necessitates[1] a rate adjustment effective March 15. Despite these outside forces, we remain dedicated to making our water as affordable as possible. A summary of rate changes can be found in the accompanying[2] attachment.

調漲原因

We are sure you understand our commitment[3] to quality. As you know, Crystal Springs is your best source of crystal-clean drinking water. We go to great lengths to supply our customers with the best nature has to offer—straight from the Diamond Range Mountains to your glass.

強調產品優勢

We will be accepting orders at the normal rate until February 28. Why not take this opportunity to put an order in now while prices are still low? You won't regret it.

招攬訂單

Drink well,

May Douglas, Customer Service Officer
Crystal Springs

Vocabulary & Phrases

1. necessitate [nɪˋsɛsə͵tet] *v.* 使成為必要
2. accompany [əˋkʌmpnɪ] *v.* 伴隨（accompanying 表示「隨同的；附上的」）
3. commitment [kəˋmɪtmənt] *n.* 承諾；保證

中文翻譯

日期：二〇一二年一月十一日
收件人：自動販賣無限公司
寄件人：m_douglas@crystalsprings.com
主旨：費用調整

致我們重視的客戶：

我們很遺憾要通知你，因為成本上漲的關係，我們無法再以同樣的低價提供新鮮泉水。以當今的經濟狀況，這種高品質的泉水要在不增加成本的情況下生產和運送是越來越不可能了。日漸上漲的油價，加上勞務費用的增加使我們必須從三月十五日起做價格調整。儘管有這些外部壓力，我們仍然致力使我們的水是大眾負擔得起的。價格變動的摘要可在隨附的檔案中找到。

我們確信你能瞭解我們對於品質的承諾。如你所知，水晶泉水是純淨飲用水的最佳來源。我們長途跋涉為客戶提供大自然的最佳獻禮——直接從鑽石山脈注入你的杯中。

在二月二十八號之前我們還是會接受平常價格的訂單。何不趁現在價格還低的時候把握機會下訂單呢？你不會後悔的。

多喝好水

客服人員　梅・道格拉斯
水晶泉水

Ⅳ. 如何用 e-mail 回覆價格調漲通知

Date: January 15, 2012
To: May Douglas
From: yolinda@v-u.com
Subject: Re: Rate Adjustment

Dear Ms. Douglas,

We appreciate the timely[1] information. Given the coming rate adjustment, we would like to take advantage of your current prices by placing an order now for 100,000 500ml bottles of your Sparkling[2] Mineral Water. Please send us an invoice[3] and we will transfer the funds to your account. Thank you.

Regards,

Yolinda Chao, Inventory Manager
Vending Unlimited

Vocabulary & Phrases

1. timely [ˋtaɪmli] *adj.*
   及時的；適時的
2. sparkling [ˋspɑrkəlɪŋ] *adj.*
   氣泡的
3. invoice [ˋɪn͵vɔɪs] *n.*
   發貨單；發票

中文翻譯

日期：二〇一二年一月十五日
收件人：梅．道格拉斯
寄件人：yolinda@v-u.com
主旨：回覆：費用調整

親愛的道格拉斯小姐：

我們很感些這個及時的資訊。假定價格即將調整，我們想利用你們現在的價格訂購十萬瓶五百毫升的氣泡礦泉水。請寄給我們一張發貨單，我們會將款項匯到妳的帳戶。謝謝。

謹上

存貨經理　曹尤琳達
自動販賣無限公司

Try it! 換你試試看!

1. 請將此信視為即將調整價格的通知。

(即將 impending)

2. 價格上漲是因為原物料成本增加了。

(原物料 raw materials)

3. 持久型的電池你找不到更好的來源了。

(持久型的 long-lasting)

4. 何不趁漲價生效前把握機會下單呢？

(生效 take effect)

答案請參閱第 373 頁

V. 用電話通知價格調漲一定要會這樣説

MP3 TRACK 112

通知價格調漲

A **I'm sorry to inform you of** a slight increase in our prices.
我很遺憾要通知你價格有些微調整。

B You don't say! How much of an increase?
不會吧！漲多少？

調漲原因

A Why are the prices going up?
為什麼漲價了？

B **It comes from rising gasoline prices.** Everything is more expensive to ship.
這是由於油價上漲。所有東西的運費都變更貴了。

強調商品優勢

A **You know as well as I do that** our whiteboard markers **are the best on the market.**
你和我都知道我們的白板筆是市面上最棒的。

B They definitely are. We sell more of your brand than any other.
確實是。和其他牌子比起來，你們的牌子我們是賣比較多。

A **As you are well aware, we are the leading name in** designer home furnishings.

如同你所熟知的，我們是設計師家居裝潢的領導品牌。

B Yes, I agree. Your brand is indeed popular with our customers.

是的，我同意。你們的牌子很受我們的客人喜愛。

招攬訂單

A **You might want to put an order in now before the rate adjustment goes into effect.**

在價格調整生效之前，你可能會想要現在下訂單。

B That would probably be a good idea.

那可能會是個好主意。

VI. 用電話通知價格調漲

頂尖建設的尼爾打電話通知老客戶價格即將全面調漲，希望他們能把握機會下單。

LISTENING 請聽 MP3 TRACK 113 ☐ | SPEAKING 請跟著 MP3 唸唸看 ☐

Neil: Hello, Sunny. It's Neil from Best Builders.

Sunny: Hi, Neil. Nice to hear from you. How've you been?

Neil: Not bad. Listen. I'm calling to follow up on an e-mail I sent you last week.

Sunny: I didn't get it. I've just got back from vacation and I haven't had time to go through all my e-mails yet. What was it about?

發出漲價通知

Neil: Unfortunately, we've had to slightly raise our prices across the board.

Sunny: Why the need for the rate adjustment?

調漲原因

Neil: Basically, it's the result of rising manufacturing costs. We've had to update our machinery and hire more workers.

Sunny: When does the increase take effect?

招攬訂單

Neil: That's why I'm calling. The end of the month will be your last chance to get our products at their current prices. I thought you might not want to miss the opportunity.

Sunny: You're right. Let me write up an order and get back to you. Thanks for calling.

Neil: Thank you. You must have a lot to do. I'll let you go. Take care.

Sunny: Thanks again. Bye now.

妳好，桑妮。我是頂尖建設的尼爾。

嗨，尼爾。很高興聽到你的消息。你最近如何？

還不錯。聽著。我打來是要追蹤我上星期寄給妳的信。

我沒有收到。我剛度假回來，還沒有時間去看我所有的電子郵件。是關於什麼事？

很遺憾，我們必須全面微調價格。

為什麼需要調整價格呢？

基本上，這是製造成本上漲的結果。我們必須更新我們的機器，並且僱用更多的工人。

調漲什麼時候生效？

這就是我打來的原因。本月底將是你們以目前價格訂購我們商品的最後機會。我想妳不會想錯失這次機會。

你說的對。我來寫一份訂單然後回寄給你。謝謝你打來。

謝謝。妳一定還有很多事要做。我不打擾了。保重。

再次感謝。再見。

Try it! 換你試試看!

WRITING 請依提示寫出完整句子	☐
LISTENING 請聽 MP3 TRACK 114	☐
SPEAKING 請跟著 MP3 唸唸看	☐

1. Ⓐ Hello Martin. What was the news that you wanted to tell me?

   Ⓑ ______________________________

   (sorry / inform / slight / rate adjustment / May 21)

   Ⓐ That's too bad.

2. Ⓐ What's the reason for the rate adjustment?

   Ⓑ ______________________________

   (result / higher shipping costs)

3. Ⓐ I'm not sure if we'll be able to sell your products if you plan to raise the prices.

   Ⓑ I guarantee you'll have no trouble selling them. ______________________________

   (our / footwear / best / market)

4. Ⓐ When will the price increase take effect?

   Ⓑ It will take effect next Tuesday. ______________________________

   (last chance / get / product / same low prices)

答案請參閱第 373 頁

UNIT
22 詢問其他付款方式
Request for Easier Payment Terms

I. 詢問其他付款方式一定要會的單字片語

MP3 TRACK 115

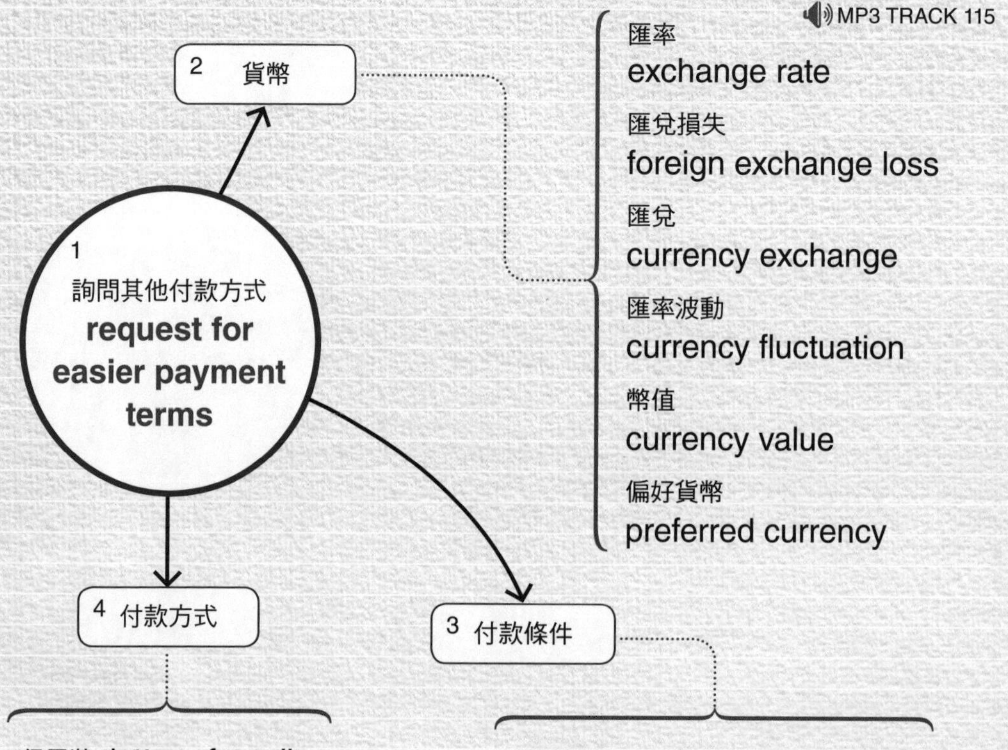

信用狀 letter of credit
銀行支票 bank check
銀行轉帳 bank transfer
信用卡 credit card
轉帳匯款 wire transfer
匯款 remittance

預付款項 up-front payment
預先付款 payment in advance
部分支付 partial payment
一次付清 full payment
頭期款 down payment
分期付款 installment payment

Ⅱ. 詢問其他付款方式一定要會的句型

句型 1 說明情況

Please allow me to explain sth.

例 Please allow me to explain our situation.
請容我解釋一下我們的情況。

句型 2 希望的付款方式

We prefer sth（付款方式）

例 For our international affairs, however, we prefer letters of credit.
然而，對我們國際事務來說，我們比較希望使用信用狀。

句型 3 等候回覆

We await . . .

例 We await your reply.
我們等候你的回覆。

句型 4 提出疑慮

We are afraid (that) + S. + V.

例 We are afraid the process may be difficult to coordinate.
我們擔心這程序或許會有協調上的困難。

Ⅲ. 如何用 e-mail 詢問其他付款方式

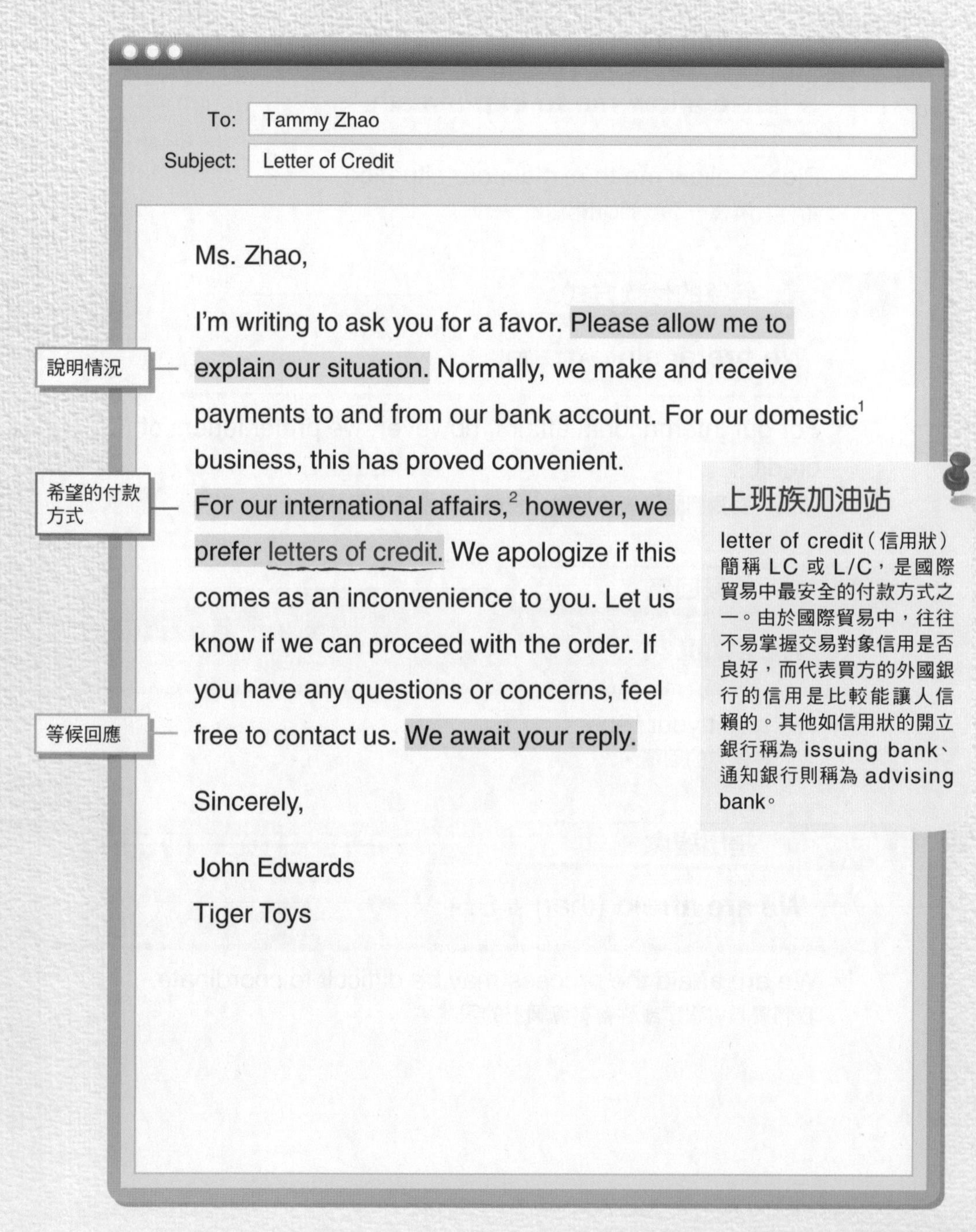

To: Tammy Zhao

Subject: Letter of Credit

Ms. Zhao,

I'm writing to ask you for a favor. Please allow me to explain our situation. Normally, we make and receive payments to and from our bank account. For our domestic[1] business, this has proved convenient. For our international affairs,[2] however, we prefer letters of credit. We apologize if this comes as an inconvenience to you. Let us know if we can proceed with the order. If you have any questions or concerns, feel free to contact us. We await your reply.

Sincerely,

John Edwards

Tiger Toys

上班族加油站

letter of credit（信用狀）簡稱 LC 或 L/C，是國際貿易中最安全的付款方式之一。由於國際貿易中，往往不易掌握交易對象信用是否良好，而代表買方的外國銀行的信用是比較能讓人信賴的。其他如信用狀的開立銀行稱為 issuing bank、通知銀行則稱為 advising bank。

Vocabulary & Phrases

1. domestic [də`mɛstɪk] *adj.*
   國內的
2. affair [ə`fɛr] *n.*
   事務（通常為複數）

中文翻譯

收件人：趙泰咪
主旨：信用狀

趙小姐：

我寫信是想要請妳幫忙。請容我解釋一下我們的情況。一般情況下，我們是從銀行帳戶來支出和收取款項。對我們國內生意往來而言，這已證明相當便利。然而，對我們國際事務來說，我們比較希望使用信用狀。如果這會對妳造成任何不便我們在此表示歉意。如果訂單可以繼續進行的話，請讓我們知道。如果妳有任何問題或疑慮，隨時都可以聯絡我們。我們等候妳的回覆。

謹上

約翰．艾德華茲
老虎玩具

文化補給站

國際企業之間常見的付款方式包括下列幾種：信用狀（Letter of Credit, L/C）、跟單託收（Documentary Collections, D/C）、記帳方式（Open Account, O/A）、PayPal 等。

以預付現金（cash-in-advance）的付款方式為例，包括銀行轉帳（wire transfer）、信用卡（credit card）、支票（check）等。這種付款機制讓出口商得以避免信貸風險（credit risk），然而買方通常不喜歡這種方式，因為可能會產生現金流（cash-flow）的問題，而且還有屆時可能無法交貨的疑慮。因此，出口商若堅持預付現金為唯一付款方式，其客戶很可能就會流失到願意提供更多彈性付款方式的廠商。

Ⅳ. 如何用 e-mail 回覆其他付款方式

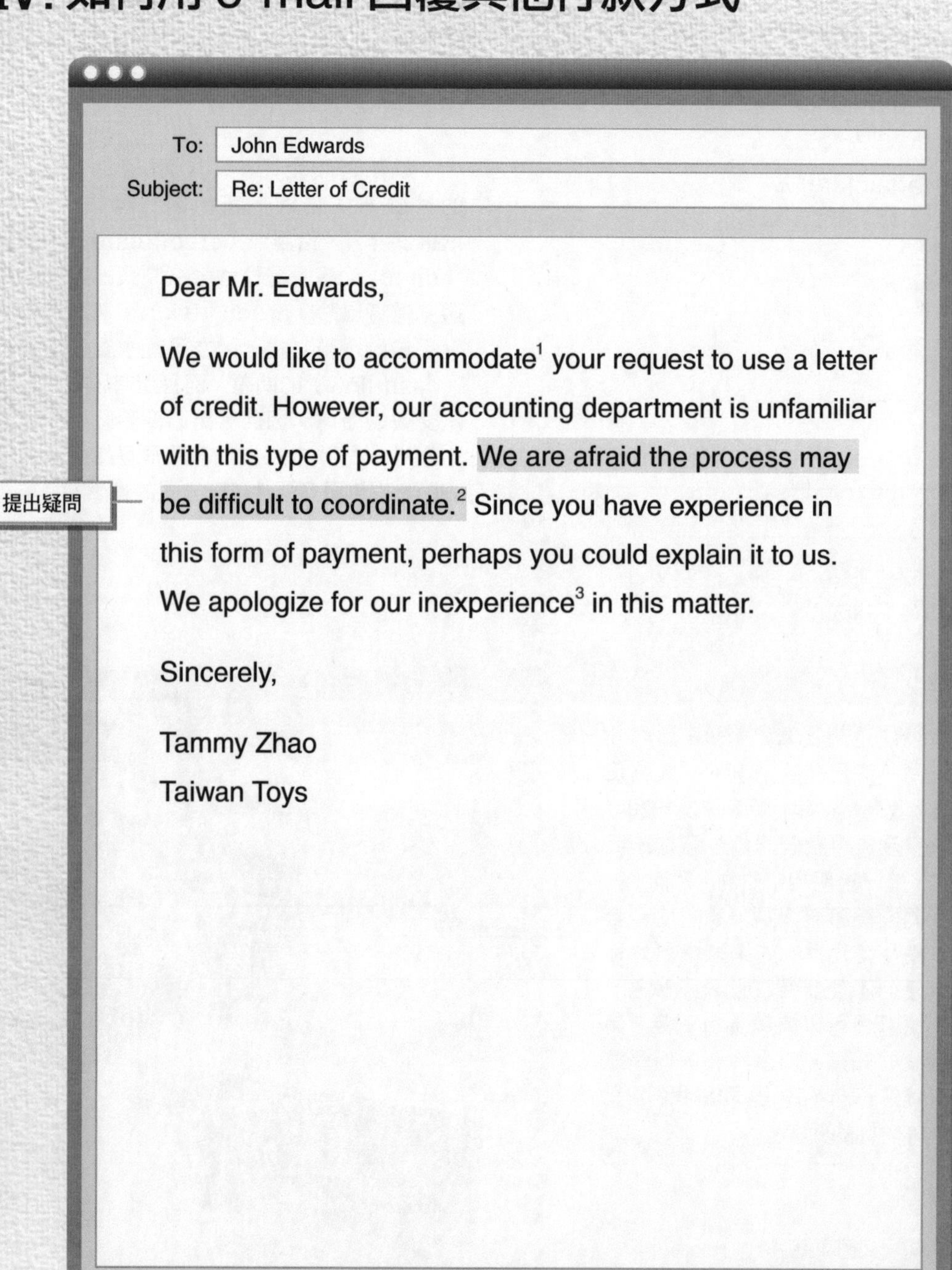

To: John Edwards

Subject: Re: Letter of Credit

Dear Mr. Edwards,

We would like to accommodate[1] your request to use a letter of credit. However, our accounting department is unfamiliar with this type of payment. We are afraid the process may be difficult to coordinate.[2] Since you have experience in this form of payment, perhaps you could explain it to us. We apologize for our inexperience[3] in this matter.

Sincerely,

Tammy Zhao

Taiwan Toys

Vocabulary & Phrases

1. accomodate [əˋkɑmə͵det] *v.*
配合；順應

2. coordinate [koˋɔrdə͵net] *v.*
協調一致

3. inexperience [͵ɪnɪkˋspɪriəns] *n.*
不熟練；無經驗

中文翻譯

收件人：約翰．艾德華茲
主旨：回覆：信用狀

親愛的艾德華茲先生：

我們願意順應你的需求使用信用狀。然而，我們的會計部門對於此種付款方式並不熟悉。我們擔心過程可能會難以協調。由於你們對此種付款方式有經驗，或許你們可以向我們說明。我們對這些事務缺乏經驗，在此表示歉意。

謹上

趙泰咪
台灣玩具

Try it! 換你試試看!

1. 我不太確定要怎麼說明我們的狀況。

____

____

____

2. 對於較大批的貨物，我們比較希望能事先取得款項。

____

____

____

(事先 in advance)

3. 我們期待能聽到你的回應。

____

____

____

4. 恐怕我們沒有資金可以預先付清。

____

____

____

答案請參閱第 373 頁

V. 電話詢問其他付款方式一定要會這樣説

MP3 TRACK 116

解釋狀況

A What seems to be the problem?

可能是怎麼樣的問題呢？

B **Frankly, I'm not sure how to describe our situation.**

坦白說，我不太曉得該怎麼描述我們的狀況。

A **Allow me to explain the circumstances we're facing.**

容我解釋一下我們正面臨的情況。

B Go ahead. I'm listening.

請說。我在聽。

希望的付款方式

A How would you like to pay for the order?

這筆訂單你想要怎麼付款？

B **When it comes to larger orders our accountants prefer bank transfers.**

若是較大筆的訂單，我們會計人員比較喜歡的是銀行匯款。

上班族小叮嚀

付款方式包括 letter of credit（信用狀）、telegraphic transfer（電匯）、pay on delivery（貨到付款）、cash against documents（憑單付現）等。

A Would it be possible to pay in installments?

有沒有可能分期付款呢？

B **We accept only full payment up front for international orders.**

國際訂單我們只接受預付全額。

等待回覆

A We are still waiting to hear back from you.

我們仍在等候你的回覆。

B Sorry about that. We've been flooded with orders lately.

很抱歉。我們最近湧近很多的訂單。

提出疑問

A **We are concerned about transfer fees.**

我們擔心轉帳費用。

B You don't have to worry about that. There are none.

那一點你不必擔心。不需要任何費用。

VI. 如何用電話詢問其他付款方式

連森鋼琴供應商打電話告知客戶他們的國際訂單無法接受分期付款。

LISTENING 請聽 MP3 TRACK 117 ☐ | SPEAKING 請跟著 MP3 唸唸看 ☐

Selina: Selina Song speaking. How may I help you?

Fred: Hi, Selina. My name is Fred Lenson from Lenson Piano Supply. I believe you have an order with us. Am I right?

Selina: That's right.

說明情況 — Fred: OK. I'm afraid we have a situation.

Selina: What seems to be the problem?

希望的付款方式 — Fred: For international orders, I'm afraid I can't accept payment in installments. You'll have to make the entire payment up front.

Selina: I understand. Are you sure there is no way we can come to an agreement on this?

Fred: Unfortunately, not. I just can't accept the risk.

Selina: That's too bad. I may have to cancel my order entirely. At the very least, I will need to postpone it.

Fred: I don't mind putting the order on hold. Let me know when you are prepared to pay, and I'll process your order right away.

Selina: Thank you, Mr. Lenson.

Fred: You're welcome. Have a good day.

我是宋瑟琳娜。有什麼可以效勞的?

嗨,瑟琳娜。我是連森鋼琴供應商的佛萊德・連森。我知道妳和我們有一筆訂單。對吧?

沒錯。

好的,恐怕我們有點狀況。

是怎麼樣的問題呢?

對國際訂單來說,我恐怕無法接受分期付款。妳必須事先付清全額。

我瞭解。你確定這件事我們沒有其他方法可以達成協議嗎?

很遺憾,沒有。我不能冒這個險。

那太糟了。我可能得將訂單全部取消了。至少,我得將訂單擱延。

我不介意暫緩這筆訂單。妳準備好方便付款的時候請讓我知道,我會立即處理妳的訂單。

謝謝你,連森先生。

不客氣。祝妳有個愉快的一天。

Try it! 換你試試看!

WRITING	請依提示寫出完整句子	☐
LISTENING	請聽 MP3 TRACK 118	☐
SPEAKING	請跟著 MP3 唸唸看	☐

1. Ⓐ Is there a problem I can help you with?

   Ⓑ ______________________________

   (not sure / explain / situation)

2. Ⓐ What kind of payment terms were you considering?

   Ⓑ ______________________________

   (this order / receive / payment / in advance)

   Ⓐ I think we can accept those terms.

3. Ⓐ I'll need to speak with my associates about this.

   Ⓑ ______________________________

   (we / hope / give / answer / soon)

   Ⓐ We'll get back to you as soon as possible.

4. Ⓐ I think the prices we've discussed are quite fair.

   Ⓑ ______________________________

   (we / concern / additional / shipping costs)

   Ⓐ There's no need to be concerned about that. We'll cover the shipping.

答案請參閱第 373 頁

UNIT 23 下訂單
Placing an Order

I. 下訂單一定要會的單字片語

MP3 TRACK 119

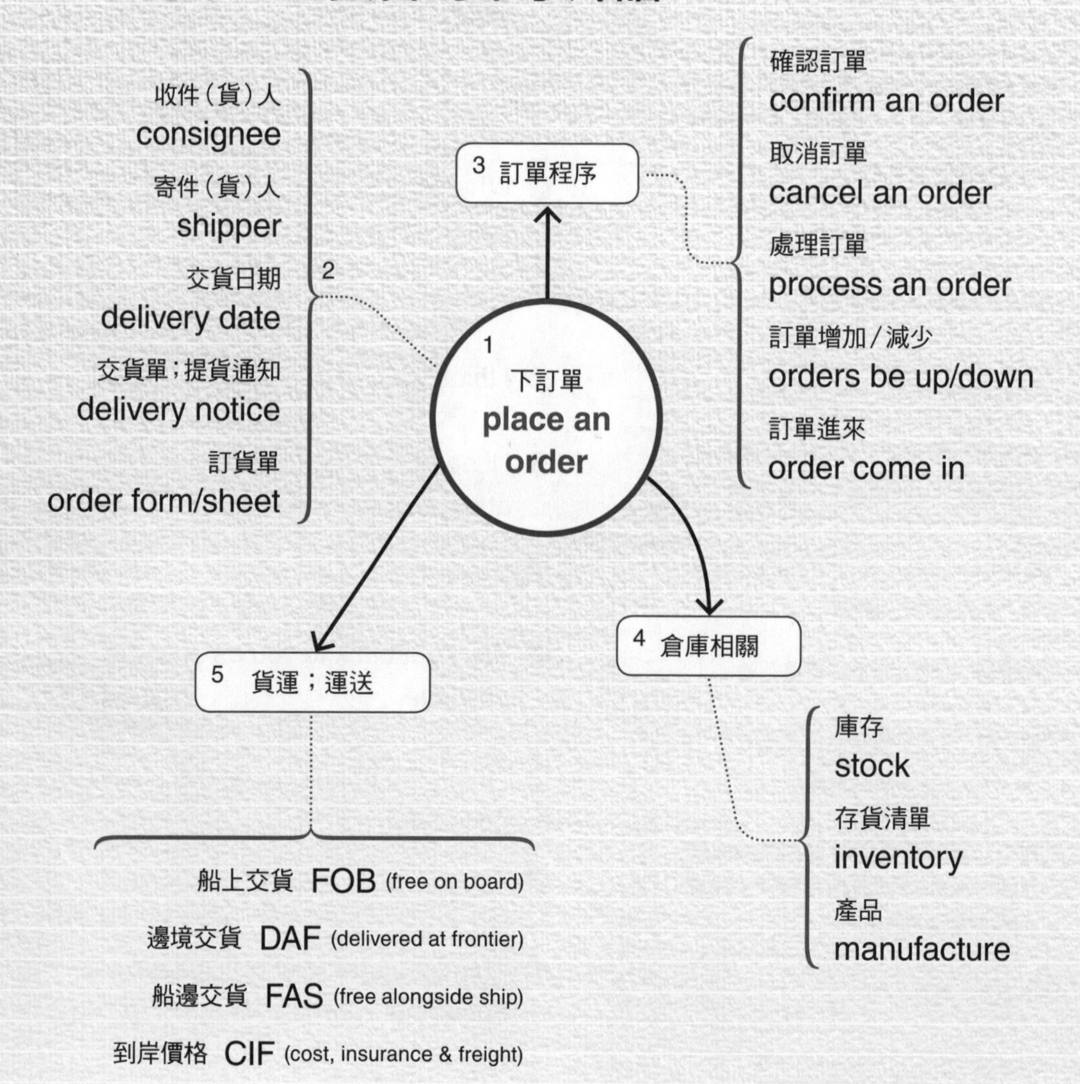

Ⅱ.下訂單一定要會的句型

句型 1 說明商品型號、規格

I'd like to place an order for sth（型號、規格等）

例 I'd like to place an order for a number of your V-5A speakers.
我想要訂購一些你們 V-5A 型的喇叭。

句型 2 說明首次訂購數量

We'd like to start with sth（數量）

例 We'd like to start with an initial order of a thousand units.
首筆訂單我們想訂購一千組。

句型 3 詢問價格

Would it be possible for you to lower the price if we ordered sth（數量）?

例 Would it be possible for you to lower the price if we ordered a thousand speakers?
如果我們訂一千個喇叭，你們有沒有可能把價格降低一點？

句型 4 送件日期

When is the earliest we can expect to receive sth（訂單）?

例 When is the earliest we can expect to receive this order?
我們最快何時可拿到這批貨呢？

Ⅲ. 如何用 e-mail 下訂單

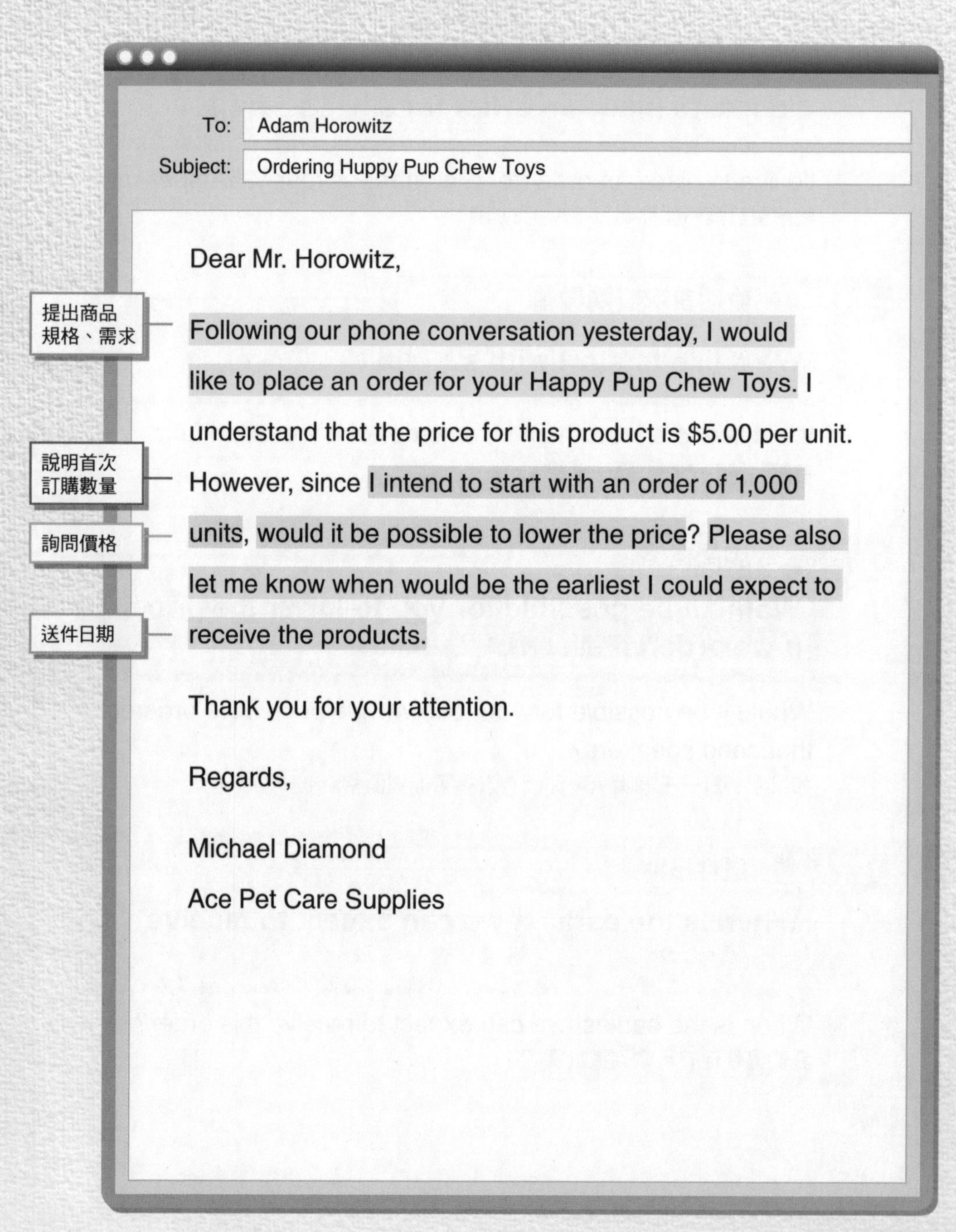
To: Adam Horowitz

Subject: Ordering Huppy Pup Chew Toys

Dear Mr. Horowitz,

Following our phone conversation yesterday, I would like to place an order for your Happy Pup Chew Toys. I understand that the price for this product is $5.00 per unit. However, since I intend to start with an order of 1,000 units, would it be possible to lower the price? Please also let me know when would be the earliest I could expect to receive the products.

Thank you for your attention.

Regards,

Michael Diamond

Ace Pet Care Supplies

中文翻譯

收件人：亞當・荷洛維茲
主旨：訂購快樂狗咀嚼玩具

親愛的荷洛維茲先生：

接續我們昨天在電話裡談論的話題，我想要訂購你們的快樂狗咀嚼玩具。我了解這商品每組價格為五塊美金。但是，因為我首批想要訂購一千組，有可能降低價格嗎？並請讓我知道最早什麼時候可以收到商品。

多謝關照。

謹上

邁可・戴蒙
首席寵物護理用品

文化補給站

看懂訂購單

訂購單（purchase order）上通常會有下列這些資訊，說明訂購商品、數量、價錢，以及交貨時間、方式等條件。以下是一些常見的用語：

Company A has this day bought the under mentioned goods from Company B to be delivered in good order subject to the terms and conditions stated hereunder, unless otherwise specified:（A公司向B公司購買下列提及的物品，除非另有註明，否則皆需按照下列陳述完好運達：）

Description: P/N（part number）	描述：產品編號
Quantity	數量
Unit Price	單價
Amount	總計
Shipping Documents	出貨文件
Destination	目的地
Delivery	交貨期
Packing	裝運
Remarks	附註

Please acknowledge receipt and acceptance of this order as soon as is convenient by returning the copy duly signed.（請以正式署名並回傳副本，儘早告知收到並接受這份訂單。）

Ⅳ. 如何用 e-mail 回覆下單

To: Michael Diamond

Subject: Re: Ordering Happy Pup Chew Toys

Dear Mr. Diamond,

Thank you for expressing your interest in Pet Plus's products. We would be happy to offer you a 10% discount from the listed price for an order of 1,000 units of our Happy Pup Chew Toys. Unfortunately, this item is currently out of stock.[1] We should, however, have it back in stock within three weeks. Our estimated[2] shipping time is 2–4 weeks.

If you would like to pre-order this product, please do not hesitate to contact me and I would be happy to assist you. My telephone number is: 123-445-5678.

Regards,

Adam Horowitz

Pet Plus

上班族加油站

打九折的英文

✗	不是 ~~90% discount~~
✓	而是 10% discount

表示折扣掉百分之十的價錢。你也可以說成 10% off。

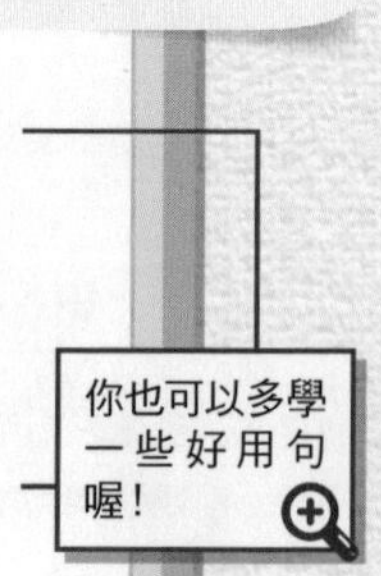

Vocabulary & Phrases

1. out of stock [stɑk] 無現貨的
2. estimate [ˋɛstə͵met] *v.* 估計

中文翻譯

收件人：邁可．戴蒙
主旨：回覆：訂購快樂狗咀嚼玩具

親愛的戴蒙先生：

感謝您對寵愛寵物的產品表示興趣。關於一千件快樂狗咀嚼玩具的訂購，我們很樂意提供您標價九折的優惠。可惜，這件商品目前缺貨中。我們將會在三個星期內補貨。我們預估的出貨時間是二至四個星期。

如果您想要預訂此商品，請不要客氣儘管與我聯繫，我將很樂意為您服務。我的電話是 123-445-5678。

謹上

亞當．荷洛維茲
寵愛寵物

Try it! 換你試試看!

1. 我想要訂購你們十六盎司的木柄鐵鎚。

(木柄鐵槌 Wood Handle Hammers)

2. 我們首批想要訂購一百件。

3. 如果我們訂購一百五十件的話，你可以降低價格嗎？

4. 大概什麼時候會收到這批貨？

答案請參閱第 373 頁

V. 用電話下訂單一定要會這樣說

MP3 TRACK 120

商品規格、需求

A	Which model did you want to order?	你想要訂哪一型？
B	**I'd like to place an order** for your V-5A speakers, please.	我想要訂購你們的V-5A 型喇叭。

訂購數量

A	**How many pieces are you looking at?**	你們打算訂多少件？
B	We'd like to start with / We were thinking about an initial order of a thousand units.	首批訂單我們想訂購一千組。

英文單位的說法除了 pieces/pcs、unit 之外還包括：bottle（瓶）、case（箱；盒）、drum（桶）、lot（批）等。

詢問庫存

A	That's quite a few.	那數目還不少。
B	**Do you have that amount available for immediate shipment?**	你們有足夠的庫存可以馬上出貨嗎？

詢問價格

A **What are you charging for** the basic model?

你們基本款的要價是多少？

B Our listed price is US$100 per unit for the basic model.

我們基本款每組的標價是一百美元。

大量訂購折扣

A **Would it be possible to lower the price if we ordered** 1,000 speakers?

如果我們訂一千個喇叭，你們有沒有可能降價？

B Unfortunately, our discounted prices start at a minimum order of 2,000 units. If you ordered 2,000 speakers, the price would be $80 per speaker.

很遺憾，我們的折扣價至少要訂購兩千組。如果你們訂購兩千個喇叭，那麼每個喇叭單價是八十美元。

送件日期

A **When is the earliest we can expect to receive this order?**

我們最快何時可拿到這批貨呢？

B I can get these out the door to you today—so, within a week.

我今天可以出貨給你——所以是一個星期以內。

VI. 如何用電話下訂單

DIY 居家維修店買方代表諾拉致電給客戶克萊夫，準備下訂單。

LISTENING 請聽 MP3 TRACK 121 ☐ | SPEAKING 請跟著 MP3 唸唸看 ☐

Nora: Clive? I'm glad I managed to catch you at your desk.

Clive: Hello, Nora. I was just going to call you about our new line of LED light bulbs.

提出商品規格、需求

Nora: Believe it or not, that's exactly why I'm calling. I'd like to place an order for those bulbs.

Clive: Great to hear it. What quantity are we looking at here?

說明訂購數量

Nora: We'd like to start with an order of a thousand cases. If they sell well, we'll increase this number down the line.

Clive: Understood. And we may be able to swing a discount for you if that happens.

詢問價格

Nora: When do you think we can receive this order?

Clive: I can get these out the door to you today—so within a week.

Nora: Perfect. We've got high hopes for these LED lights of yours, Clive.

Clive: They won't let you down, Nora.

克萊夫嗎？真高興我總算找到你了。

嗨，諾拉。我才正要打電話跟妳說我們新的 LED 燈泡系列。

不蓋你，這正是我打電話來的原因。我想訂購那些燈泡。

很高興聽妳這麼說。這次考慮的數量是多少呢？

首筆訂單我們想訂購一千箱。如果銷路不錯，未來我們會增加數量。

瞭解了。如果訂單增加的話，我們有可能幫妳爭取到折扣。

你認為我們何時可拿到這批貨呢？

我今天可以出貨給妳——所以是一個星期以內。

太棒了。我們對你們公司的 LED 燈泡寄予厚望喔，克萊夫。

諾拉，它們不會讓妳失望的。

Try it! 換你試試看!

WRITING 請依提示寫出完整句子	☐
LISTENING 請聽 MP3 TRACK 122	☐
SPEAKING 請跟著 MP3 唸唸看	☐

1. A Which model did you want to order?

   B ______________________________

   ____

   (I / A-51 office chairs)

2. A How many pieces are you looking at?

   B ______________________________

   ____

   (we / wanted / place an order / 30 office chairs)

3. A The total cost will be $55,000.

   B ______________________________

   ____

   (would / discount / fifty units)

4. A You'll receive your order very soon.

   B ______________________________

   ____

   (could / tell / earliest / receive/ order)

   A We'll be able to deliver it in two weeks.

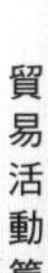

貿易活動篇

答案請參閱第 373 頁

UNIT
24 安排出貨
Arranging Shipment

I. 安排出貨一定要會的單字片語

MP3 TRACK 123

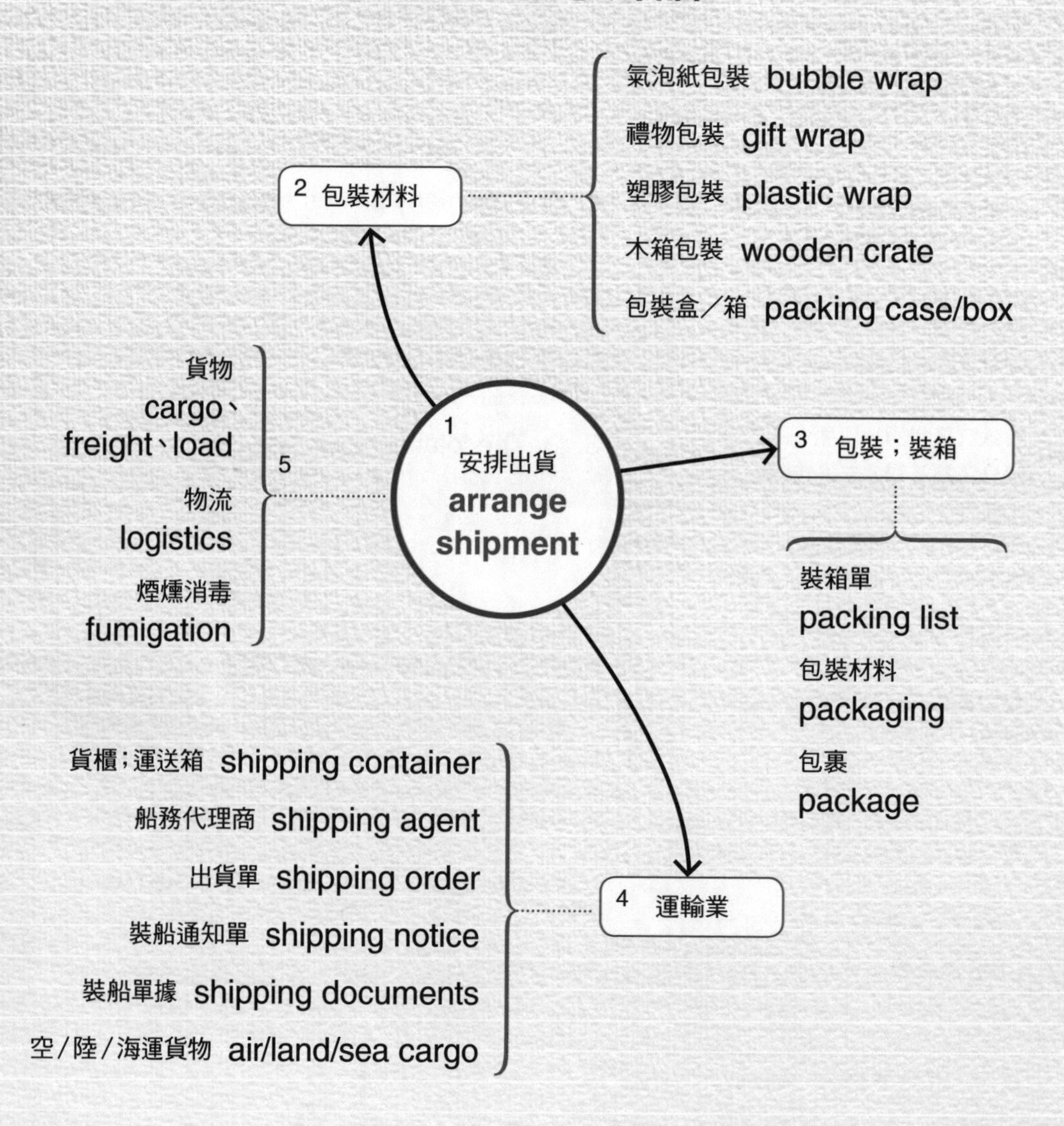

Ⅱ. 安排出貨一定要會的句型

句型1 提出問題

We still have a few questions about sth.

例 We still have a few questions about some of the details.
我們對某些細節還有一些問題。

句型2 運送日期

. . . **provide us with an estimated delivery date.**

例 We hope you can provide us with an estimated delivery date.
我們希望你可以提供我們一個預計的交貨日期。

句型3 出貨確認

Will we be getting a shipping confirmation . . .?

例 Will we be getting a shipping confirmation once our order leaves port?
我們的貨物一離港時，我們會拿到出貨確認單嗎？

句型4 包裝需求

Would it be possible to carefully wrap each unit in + 包裝需求？

例 Would it be possible to carefully wrap each unit in bubble wrap?
可以每組都用氣泡紙仔細包裝嗎？

Ⅲ. 如何用 e-mail 安排出貨

To: Valerie Monroe

Subject: Shipping Arrangements

Dear Ms. Monroe,

We know that your company will be covering the cost of shipping for which we are grateful. We still have a few questions about some of the details, however. We are curious as to which shipping company you use. We would also like to know whether we will get a tracking[1] number associated with our order. In addition, we hope you can provide us with an estimated delivery date. We may need to expedite[2] our order. Finally, will we be getting a shipping confirmation once our order leaves port?

提出問題

運送日期

運送通知

We also have a special packaging request to ask of you. Would it be possible to carefully wrap each unit in bubble wrap? We've had orders from other companies get damaged in transit[3] and would like to avoid this happening again.

包裝需求

Thank you very much for fielding so many specific questions. We greatly appreciate your attention to our order.

Sincerely,

Yvonne Chen

Asia Pacific Producers, Inc.

上班族加油站

field 在此指的是回答難以應付的問題，也可用 cope with、address 來表示「應付」的意思。

Vocabulary & Phrases

1. track [træk] *v.* 追蹤
2. expedite [ˋɛkspə͵daɪt] *v.* 迅速執行
3. transit [ˋtrænsət] *n.* 運送

中文翻譯

收件人：瓦樂莉・門羅
主旨：貨運安排

親愛的門羅小姐：

我們得知貴公司會負擔運費，對此我們深表感謝。然而，我們對某些細節還有一些問題。我們好奇你們使用的是哪一間貨運公司。我們也想知道我們是否會取得訂單的追蹤號碼。除此之外，我希望妳可以提供我們一個預計的交貨日期。這筆訂單我們可能會需要加快進行。最後，我們的貨物一離港時，我們會拿到出貨確認單嗎？

我們也有一些特殊的包裝要求要問妳。可以每組都用氣泡紙仔細包裝嗎？我們從其他公司訂購的貨物曾經在運送過程中受到損壞，希望能避免再發生此事。

非常感激妳回答這麼多細節問題。我們很感謝妳對我們貨物的關注。

謹上

陳依芳
太平洋亞洲製造有限公司

Try it! 換你試試看!

1. 對於運送細節我們有一些疑慮。

2. 你可以提供我們一個預估的寄送時間嗎？

3. 我們會拿到出貨確認單嗎？

4. 要求禮品包裝會不會太多了？

答案請參閱第 374 頁

Ⅳ. 如何用 e-mail 回覆出貨安排

To: Yvonne Chen

Subject: Re: Shipping Arrangements

Dear Ms. Chen,

We are more than happy to answer your questions.

Intercontinental Freight[1] handles our shipping. They are an internationally ranked shipping company with a five-star rating. You can be assured that they will take good care of your order. A tracking number is indeed supplied for all our shipments. You will find it on the Shipping Confirmation which will be e-mailed to you promptly after your order leaves our warehouse. Once we ship your order, it should take approximately two weeks to reach you. We could ship as early as tomorrow, providing you have no further concerns. That will make the earliest arrival time August 9. I trust that meets with your approval?

As for your special request, you should know that we wrap all sensitive[2] machine parts in bubble wrap and fill each container with Styrofoam pellets. We also use a very durable[3] cardboard for our shipping containers. We have never had any problems with units damaged in shipping. Don't worry. Your order is in good hands.

上班族加油站

be in good hands 表示受到良好的照顧，而 be in sb's hands 表示某事是某人的職責，

It is our pleasure to serve you. If everything is to your liking, we can ship your order tomorrow. We await your reply.

Sincerely,

Valerie Monroe

Lincoln Machinery

Vocabulary & Phrases

1. freight [fret] *n.* 貨物；貨運
2. sensitive [ˋsɛnsətɪv] *adj.* 易受損的
3. durable [ˋdurəbḷ] *adj.* 耐用的

中文翻譯

收件人：陳依芳
主旨：回覆：裝運安排

親愛的陳小姐：

我們很樂意回答妳的問題。

洲際貨運負責我們的運送。他們是列入國際排名的公司，有五顆星的等級。妳可以放心，他們一定會妥善處理妳的訂貨。事實上我們所有的貨物都有追蹤號碼。妳可以在運送確認單上找到，確認單會在訂貨離開我們倉庫後以電子郵件的方式立即傳送給妳。一旦我們將訂單裝貨運送，應該約莫兩個星期就會送達妳手上。倘若妳沒有其他疑慮的話，我們最早可以在明天出貨。這樣的話最早可以在八月九日抵達。我相信妳會贊同的，是嗎？

至於妳的特殊需求，妳應該知道我們會以氣泡紙仔細包裝所有靈敏的機械零件，並且將每個裝箱都填入保麗龍球。我們的運送箱也採用非常耐用的硬紙板。我們從未在運送過程中出現組件受損的問題。別擔心。妳的訂貨會受到妥善處理。

能替妳服務是我們的榮幸。如果每件事都合妳意的話，我們明天就可以出貨。等候妳的答覆。

謹上

瓦樂莉‧門羅

林肯機械

V. 用電話安排出貨一定要會這樣説

MP3 TRACK 124

✆ 貨運問題

A **Could we clarify some details in the shipping arrangements?**
我們可以釐清貨運安排上的一些細節嗎？

B Certainly. Which part of the shipping were you concerned about?
當然。你們擔心運送的哪個部份？

✆ 詢問出貨通知確認

A **Will we be getting an e-mail to confirm that our shipment has been sent out?**
我們會收到電子郵件確認貨物已經寄出了嗎？

B We can provide that if you require it.
如果你們要求的話我們可以提供。

A **We would greatly appreciate it.**
我們會很感謝。

B I'll see that you get one.
我會確認你有收到。

詢問到達日期

A **When can we expect our shipment to arrive?**

我們能預期貨物什麼時候抵達嗎？

B It should be there no later than May 10.

應該不會晚於五月十日。

回應出貨通知

A **Will you be sending us a shipment confirmation?**

你會寄出貨確認通知給我們嗎？

B We will e-mail you a proof of shipment once the items have left our premises.

一旦商品離開廠房我們就會以電子郵件寄出貨證明給你。

包裝需求

A **Would it be too much if we asked for nonstandard packaging?**

如果我們要求非標準包裝會不會太過分？

B That depends on the nature of the packaging you require.

那要看你們要求的是怎樣的包裝。

包裝方式還包括 pallet wrap（棧板包膜）、strapping（綁紮）、void fill（緩衝包材）等。

VI. 如何用電話安排出貨

帝王音樂的妮娜打電話給雷．維卡，詢問出貨的一些事宜。

LISTENING 請聽 MP3 TRACK 125 ☐ SPEAKING 請跟著 MP3 唸唸看 ☐

Ray: Hello. Ray Vicar speaking. How may I help you?

Nina: Hi, Ray. This is Nina from Emperor Music in Taiwan. I'm calling about our pending order. I wanted to discuss some of the shipping details with you.

提出問題

Ray: Sure. Where should we start?

Nina: First of all, we would like to know if the price of shipping includes insurance and, if so, how much.

Ray: It does. Our freight company insures the entire cost of the shipment.

運送日期

Nina: Got it. Thank you. Next, when can we expect the CDs to arrive?

Ray: They're going by air, so if we ship them on Monday, they should get there within three business days. That means Thursday, or Friday at the latest.

Nina: Excellent. So far, so good. Now I have a special request to ask of you.

Ray: I'm listening.

Nina: It's about the cases for the CDs. Could you possibly ship them in thick jewel cases and not the thin ones?

包裝需求

Ray: We can do it, but it will delay shipment by at least a week.

Nina: Not a problem. Thanks a lot.

Ray: My pleasure. Was there anything else?

Nina: No, that's it. Thanks again.

妳好。我是雷·維卡。我可以怎麼幫妳？

嗨，雷。我是從台灣帝王音樂打來的妮娜。我打來是關於一筆尚未敲定的訂單。我想要跟你談談一些運送細節。

當然。我們從哪裡開始？

首先，我們想要知道運費是否包括保險，如果有的話又是多少？

有的。我們的貨運公司有為貨物的總價值投保。

瞭解了。謝謝。接著，我們可以預期這些 CD 什麼時候到？

它們是走空運，所以如果我們星期一出貨的話，應該會在三個工作天之內到達。也就是星期四或最晚星期五。

太棒了。目前為止都很好。現在我想向你提出一個特殊要求。

請說。

是關於 CD 的盒子。可不可能用厚的 CD 硬殼出貨而不是用薄的那種嗎？

我們可以辦到，但是出貨至少要晚一個星期。

沒關係。非常謝謝。

這是我的榮幸。還有其他事嗎？

沒有，就這樣。再次感謝。

Try it! 換你試試看!

WRITING 請依提示寫出完整句子	☐
LISTENING 請聽 MP3 TRACK 126	☐
SPEAKING 請跟著 MP3 唸唸看	☐

1. Ⓐ Which part of your pending order did you want to talk about?

   Ⓑ ______________________________

   (I / discuss / shipping arrangements)

   Ⓐ Of course. What's on your mind?

2. Ⓐ Your shipment is scheduled to leave our warehouse next week.

   Ⓑ ______________________________

   (expected / delivery date / our order)

   Ⓐ It should be there by August 22.

3. Ⓐ Global Trend Shipping Company will be picking up your goods from our warehouse this Friday.

   Ⓑ ______________________________

   (we / get / shipping confirmation)

   Ⓐ Yes. We send notices out for all our shipments.

4. Ⓐ One other question.

   Ⓑ I'm all ears.

   Ⓐ ______________________________

   (possible / ship / units / individually)

   Ⓑ I'm afraid we can't do that.

答案請參閱第 374 頁

UNIT

25 催款

Request for Payment

I. 催款一定要會的單字片語

MP3 TRACK 127

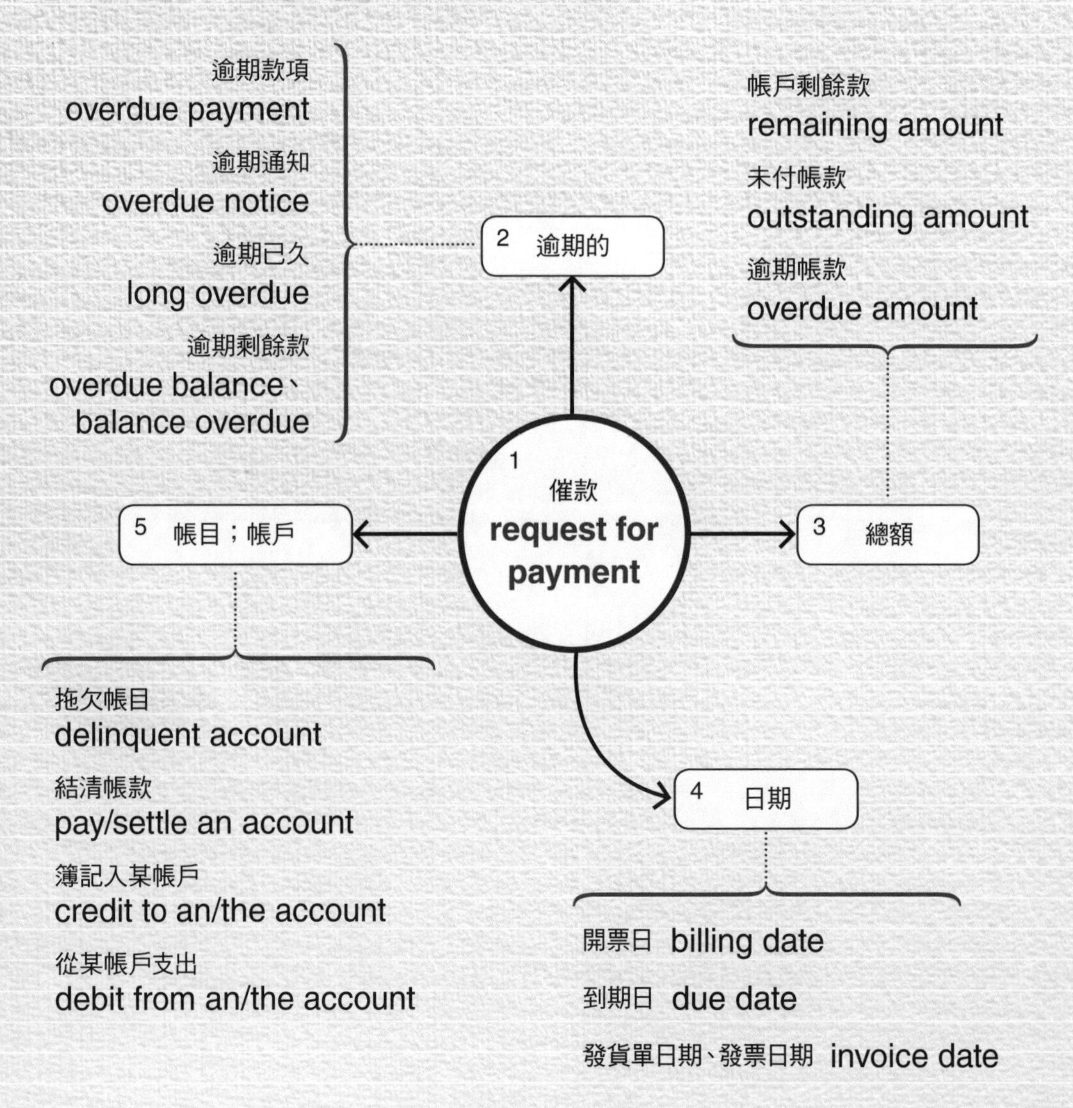

Ⅱ. 催款一定要會的句型

句型 1　通知繳款到期

This letter is to inform you that payment is now + 時間 overdue.

例 This letter is to inform you that payment is now three days overdue.
這封信是要通知你們款項已經逾期三天了。

句型 2　提醒日期及時間

Your payment of + 金額 was due on + 日期

例 Your payment of $5,356.89 was due on November 25.
你們的款項五千三百五十六點八九美元在十一月二十五日到期。

句型 3　瞭解原因

If you are unable + to V., perhaps you could explain . . .

例 If you are unable to make payment immediately or require special arrangements, perhaps you could explain the problem to us.
如果你們無法立即付款或需要特殊安排，或許可以向我們說明問題。

句型 4　要求回覆

We would appreciate hearing back from you . . .

例 We would appreciate hearing back from you by phone or e-mail at your earliest convenience.
若能儘早透過電話或電子郵件得到你們的回應，則不勝感激。

Ⅲ. 如何用 e-mail 催款

To: Hearty Health (tom_trust@heartyhealth.com)
Subject: Payment Reminder

Dear Mr. Trust,

通知繳款到期 — This letter is to inform you that your payment is now three days overdue. This regards your order of October 24 for 20 crates of multivitamins which was shipped on October 26 and delivered on November 18. 提醒日期及時間 — Your payment of $5,356.89 was due on November 25. Please remit funds to our account for the balance mentioned above.

We understand that it is sometimes difficult to make payment on time. 瞭解原因 — If you are unable to make payment immediately or require special arrangements, perhaps you could explain the problem to us. Our accounting office has experience in solving difficult payment issues. We are happy to assist our customers whatever their needs may be.

要求回覆 — We would appreciate hearing back from you by phone or e-mail at your earliest convenience. Thank you kindly.

Sincerely,

Jacky Fu
Vital Vitamins
886-2-2234-5678

中文翻譯

收件人：康健（tom_trust@heartyhealth.com）
主旨：付款通知

親愛的楚斯特先生：

這封信是要通知你的款項已經逾期三天了。這是有關你們在十月二十四日所訂購二十個貨櫃的綜合維他命，已於十月二十六日出貨並在十一月十八日到貨。你們的款項五千三百五十六點八九美元在十一月二十五日到期。請將上述的款項匯到我們的帳戶。

我們理解按時繳交款項有時候並不容易。如果你們無法立即付款或是需要特殊安排，或許可以向我們說明問題。我們的會計部門對於解決棘手的付款問題很有經驗。不管客戶的需求是什麼，我們都很樂意提供協助。

若能儘早透過電話或電子郵件得到你們的回應，則不勝感激。衷心感謝你。

謹上

傅傑克
活力維他命
886-2-2234-5678

文化補給站

寫信向客戶或客人催款時，立場要清楚，語氣要堅定且專業，最不樂見的情況是你寫了一、兩封無效又讓對方覺得受到攻擊的催款信。如果信中的語氣過於刻薄刺耳，客戶可能會覺得沒有必要繼續維持生意關係，一旦發生這樣的情況，就有可能發生客戶不付款的風險，因為他已決心不再和你有任何生意往來。

如果你寄出催款信且過了三十天未獲回應的話，建議你最好打個電話，不過與書信往來一樣，絕不口出惡言。

如果又過了一段時間對方仍未支付款項的話，接下來催收信的語氣則需要更為堅定，最終有可能得交付催款單位來處理。

當你試了所有平和的手段仍無效時，才建議你採取最不得已的手段，寄出最後通牒。另外，絕對不要在催款信件中提及要訴諸法律途徑，除非你已打定主意要這麼做。

IV. 如何用 e-mail 回覆催款信

To: Jacky Fu (Vital Vitamins)

Subject: Re: Payment Reminder

Dear Mr. Fu,

Please accept my personal apologies for failing to make payment on time. I should have realized that the Thanksgiving holiday meant I needed to send payment earlier. Our offices were closed November 24 to the 27. Nevertheless, I should have anticipated[1] this problem and sent payment on Wednesday at the latest.[2] I hope we have not lost your trust. We will transfer the entire balance to your account immediately. We are sorry for the delay. Thank you for understanding.

Sincerely,

Tom Trust

Hearty Health

上班族加油站

transfer 表示「轉讓；轉移」的意思，可作名詞和動詞用，transfer the entire balance to your account 是指「將餘額轉帳到你的帳戶」的意思。另外 bank transfer 則指「銀行轉帳」，如 payment will be made by bank transfer（款項將以銀行轉帳方式來進行）。

Vocabulary & Phrases

1. anticipate [æn`tɪsə͵pet] *v.* 預想；預料
2. at the latest *phr.* 最遲

中文翻譯

收件人：傅傑克（活力維他命）
主旨：回覆：付款通知

親愛的傅先生：

對於沒有準時付款，請接受我個人道歉。我應該要知道感恩節假期表示我必須早點將款項寄出。我們十一月二十四日至二十七日期間不上班。儘管如此，我應該事先想到這個問題，並最晚在星期三寄出款項。我希望我們沒有失去你的信任。我們會立即將所有的剩餘款項匯到你們的戶頭。我很抱歉造成延遲。感謝你的諒解。

謹上

湯姆．楚斯特
康健

Try it! 換你試試看!

1. 你們的訂單款項現在已經逾期一星期了。

____

____

____

(逾期 overdue)

2. 你們新台幣三千五百六十三元的款項在八月七日到期。

____

____

____

3. 如果你向我們解釋問題，我們將很樂意為你們解決問題。

____

____

____

(解決 resolve)

4. 我們感謝你立即回覆。

____

____

____

(立即的 prompt)

答案請參閱第 374 頁

V. 電話催款一定要會這樣說

MP3 TRACK 128

催款

A **Did you know that we haven't received payment for your latest invoice?**

你知道我們沒有收到你們最近一筆發貨單上的款項嗎？

B No, I wasn't aware of that. Let me look into that for you.

我不曉得這件事。讓我幫你查一查。

提醒款項及日期

A **Your total of** $43,789 **was due** a week ago today.

你們的總金額四萬三千七百八十九元的付款在一個星期前到期了。

B That long?! We'll get that out to you as soon as possible.

有那麼久？我們會儘早給你。

A When was the payment due?

款項是什麼時候到期？

B **It was due** this past Monday which was April 30.

是在上個星期一，四月三十日到期。

詢問延遲原因

A **Would you mind telling us the problem that led to the delay?**

你介意告訴我們是什麼問題導致延遲的嗎？

B It seems our accountant provided the wrong information to the bank.

好像是我們的會計人員提供給銀行的資訊是錯的。

希望儘快得到答覆

A **We would appreciate an answer as soon as possible.**

如蒙儘早回覆將不勝感激。

B I'll find out and get back to you right away.

我會查明並立即回覆你。

VI. 如何用電話催款

創意室內裝潢的亞當打電話給客戶通知繳款逾期，並試著瞭解逾期的原因。

LISTENING 請聽 MP3 TRACK 129 ☐ | SPEAKING 請跟著 MP3 唸唸看 ☐

Adam: Hello, Laura. This is Adam from Creative Décor.

Laura: Hi, Adam. Thanks for sending us those statuettes. What can I do for you?

通知繳款到期 — Adam: Actually, that order is why I'm calling. We haven't received your payment yet.

Laura: I had no idea. How long is it overdue?

提醒日期及時間 — Adam: Almost a week now.

Laura: My apologies, Adam. I'll get right on it.

瞭解原因 — Adam: Do you have any idea how this could have happened?

Laura: Not yet. I'm as surprised as you are. I'll definitely be having a word with my secretary.

Adam: Thanks, Laura. I appreciate your taking care of it.

Laura: I'll make sure you have the payment ASAP. I'm sorry for the trouble.

Adam: It's OK. I'll let you go now, so you can work things out. Thanks again.

妳好，蘿拉。我是創意室內裝潢的亞當。

嗨，亞當。感謝你寄給我們那些雕像裝飾。我可以為你做什麼嗎？

其實，我打來就是為了那筆訂單。我們還沒收到你們的付款。

我不清楚。逾期多久了？

到現在差不多一星期了。

亞當，我很抱歉。我會馬上處理。

妳知道為什麼會發生這種事嗎？

還不曉得。我和你一樣驚訝。我一定會和我的祕書談一談。

謝謝妳，蘿拉。我很感謝妳要處理這件事。

我會確認你儘早收到款項。我很抱歉造成你的麻煩。

沒關係。我要掛了，這樣妳才可以處理這件事。再次感謝。

Try it! 換你試試看!

WRITING 請依提示寫出完整句子	☐
LISTENING 請聽 MP3 TRACK 130	☐
SPEAKING 請跟著 MP3 唸唸看	☐

1. Ⓐ Is there a problem with my account?

   Ⓑ ______________________________

   ____

   (were / aware / payment / overdue?)

   Ⓐ I wasn't aware of that. Thank you for informing me.

2. Ⓐ Can you remind me what the total amount was?

   Ⓑ ______________________________

   ____

   (your / NT$7,800 / due / December 21)

3. Ⓐ I sincerely apologize about our lateness in settling our account.

   Ⓑ ______________________________

   ____

   (we / appreciate / explanation / delay / payment)

   Ⓐ It seems we have insufficient funds at the moment.

4. Ⓐ We will resolve the issue immediately.

   Ⓑ Thank you. ______________________________

   ____

   (we await / prompt attention / matter)

答案請參閱第 374 頁

售後處理篇

貿易往來難免要處理售後的問題，無論是處理退貨、賠償的問題，或要求技術支援，這些都是相當實用的議題，不論是向客戶說明原因以取得對方諒解，或是在維護公司權益下又不傷害彼此貿易關係，在這個篇章中我們要告訴你各種應對的技巧。

UNIT 26 更改訂單

Changing an Order

I. 更改訂單一定要會的單字片語

MP3 TRACK 131

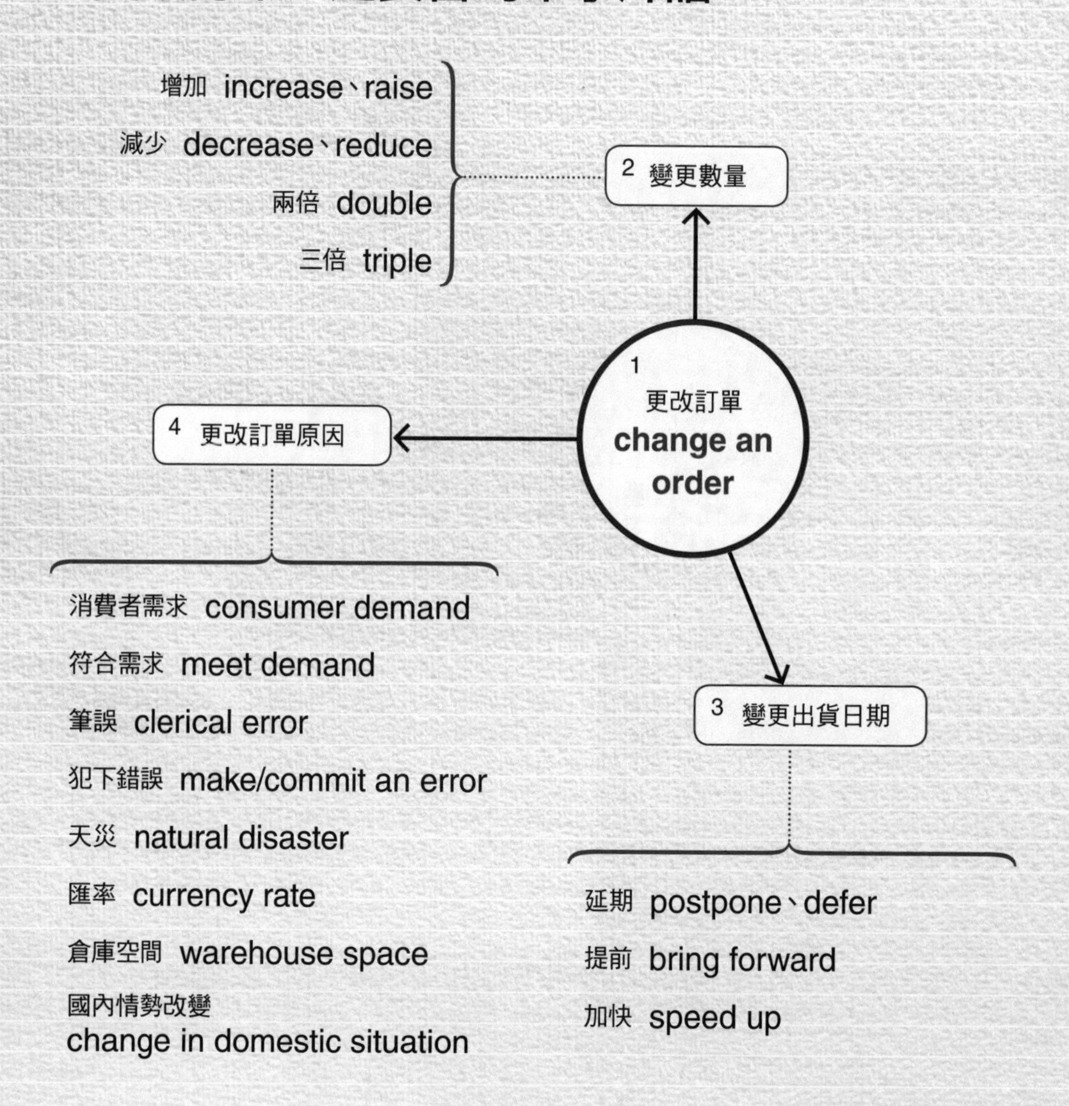

Ⅱ. 更改訂單一定要會的單字片語

句型 1 是否能更改訂單

We would like + to V. **if at all possible.**

例 We would like to increase our order if at all possible.
如果可能的話我們想要增加訂單量。

句型 2 更改原因

This is due to sth.

例 This is due to new market analyses which suggest a stronger demand than we at first anticipated.
這是因為新的市場分析顯示需求比我們原先預期的還要更強烈。

句型 3 詢問價錢

Please inform us of sth.

例 Please inform us of the new price.
請告知我們最新價錢。

句型 4 告知好消息

I'm pleased to inform you that + S. + V.

例 I'm pleased to inform you that we caught the shipment in time, though just barely.
我很高興通知你我們及時將貨品攔下，只差了一步。

Ⅲ. 如何用 e-mail 更改訂單

To: Allen Wilson

Subject: URGENT: Order Change

Dear Mr. Wilson,

I realize we have already finalized[1] the order (invoice #02318), and you may have already shipped it. However, it has come to our attention that we will need more of the same item. We would like to increase our order if at all possible. This is due to new market analyses which suggest a stronger demand than we at first anticipated. As such, could you double our order to 100,000 units? Everything else on the order remains the same.

是否能更改訂單

更改原因

You said something during our negotiations about a discount for larger orders. I can't remember now what the minimum[2] order was to obtain a lower rate. Does our order qualify now? Please inform us of the new price either way. We apologize for the inconvenience. Thank you.

詢問價錢

Sincerely,

Ethan Chen

Asia-Pacific Conglomerates,[3] Inc.

上班族加油站

sth comes to sb's attention 通常表示注意到某個問題。

Vocabulary & Phrases

1. finalize [ˋfaɪnə͵laɪz] *v.* 結束；完成
2. minimum [ˋmɪnəməm] *adj.* 最少的
3. conglomerate [kənˋglɑmrət] *n.* 企業集團

中文翻譯

收件人：艾倫・威爾森
主旨：緊急：更改訂單

親愛的威爾森先生：

我知道我們已經完成這份訂單（發貨單 02318 號），而你可能已經出貨。然而，我們剛注意到我們需要更多相同的商品。如果可能的話我們想要增加訂單量。這是因為新的市場分析，顯示需求比我們原先預期的還要更強烈。如上所述，你可以將我們的訂量增加一倍到十萬組嗎？訂單上的其他條件維持不變。

在我們議價的過程中你有提過大量訂購的折扣。我現在不記得符合較低價錢的最少訂量是多少。我們現在的訂單有符合條件了嗎？用哪種方式都可以，請告知我們最新的價錢。我們對於造成不便感到抱歉。謝謝你。

謹上

陳伊森
亞洲太平洋企業集團

Ⅳ. 如何用 e-mail 回覆訂單修改

To: Ethan Chen

Subject: Re: URGENT: Order Change

Dear Mr. Chen,

要求回覆

I'm pleased to inform you that we caught the shipment in time, though just barely.[1] Our driver was just about to leave the warehouse with your order! Fortunately, we stopped him and filled his truck with the double order you asked for. To answer your question: Yes, your order now qualifies for the 10% discount we discussed earlier. We will apply[2] it to the new invoice (#02319) which you can find attached to this e-mail. Your total is now US$80,000.

Thank you for your business and your increased order.

Sincerely,

Allen Wilson

Great Things

Vocabulary & Phrases

1. barely [ˋbɛrli] *adv.* 勉強地
2. apply [əˋplaɪ] *v.* 使適用

中文翻譯

收件人：陳伊森
主旨：回覆：緊急：更改訂單

親愛的陳先生：

我很高興通知你我們及時將貨品攔下，只差了一步。我們司機剛好要將你訂的貨運離倉庫！幸好，我們攔住他，並將你們要求的兩倍訂單上貨。回覆你的問題：是的，你的訂單現在符合我們先前提過的九折優惠。我們會在新的發貨單（#02319）上做折扣，你可以在此封信件的附檔中找到。現在總價是八萬美金。

感謝此次交易以及所追加的訂單。

謹上

艾倫·威爾森
絕妙好事

Try it! 換你試試看！

1. 我們訂單有可能減少三百組嗎？

(減少 decrease)

2. 這是會計差錯的結果。

(會計差錯 accounting error)

3. 請你一算好就告訴我們新的價錢。

(一……就…… as soon as)

4. 我很高興通知你們現在訂單已經符合折扣條件。

(符合 qualify for)

答案請參閱第 374 頁

V. 用電話更改訂單一定要會這樣說

MP3 TRACK 132

確認是否能更改訂單

A **We would like to add a few items to our order if that's OK.**

如果沒問題的話，我們想要在訂單中增加一些品項。

B Certainly. We can just draw up a new invoice for you.

當然。我們可以再擬一份新的發貨單給你。

詢問更改內容

A We need to make some changes to the order.

我們需要將訂單做一些修改。

B **What sort of changes did you have in mind?**

你想要做什麼樣的修改？

詢問修改原因

A **Why the change of heart?**

為什麼改變心意？

B No change of heart. It's actually because of a clerical error on our part.

沒有改變心意。其實是因為我們這邊筆誤。

說明修改原因

A **The change is due to an oversight in planning which we've just now caught.**

修改是因為在計畫時有所疏失，我們現在才注意到。

B That's not a problem. Let's write you up a new order.

沒問題。我們再幫你寫一份新的訂單。

確認價錢

A **How will this change affect the price?**

這項更動會影響價錢嗎？

B The rate should stay the same.

價錢應該會維持不變。

同意訂單修改

A **I'm happy to tell you that** we have your additionally requested items in stock.

很高興告訴你，你們另外要求的商品我們有庫存。

B That's wonderful news. Thank you so much.

真是個好消息。非常感謝你。

VI. 如何用電話更改訂單

台北進口公司的菲歐娜打電話詢問廠商是否能增加訂單，以及是否有額外折扣。

LISTENING 請聽 MP3 TRACK 133 ☐ | SPEAKING 請跟著 MP3 唸唸看 ☐

Fiona: Hello, Mr. Twain? This is Fiona Sun from Taipei Imports in Taiwan. We have an active order with you.

Eddie: Hello, Ms. Sun. Let me pull up your information. Could you give me the number on the invoice?

Fiona: Of course. I have it right here. It's 10498.

Eddie: OK. Here it is. You ordered 1,000 sheets of leather, is that right?

是否能更改訂單 — Fiona: Correct. But we'd like to change that order to 2,000 sheets if that's OK.

詢問更改原因 — Eddie: That should be alright. We haven't shipped your order out yet. Any reason for the sudden change?

更改原因 — Fiona: It seems we can save money on shipping if we order more.

Eddie: That's always good.

詢問價錢 — Fiona: Indeed. I presume this shouldn't affect the price at all? Or would you like to give us a discount by any chance?

Eddie: I'm sorry. We've actually already given you the discounted rate.

Fiona: I understand. Never hurts to ask, though. Well then, thanks for the help, Mr. Twain. It's been a pleasure.

Eddie: Thank you, Ms. Sun. We'll get this shipment out to you right away.

上班族加油站

在商業電話禮儀中，為了感謝別人的協助，在結束電話前可以跟對方說 it's been a pleasure，或是 it's been a pleasure talking to you。

你好？吐溫先生？我是從台灣的台北進口公司打來的孫菲歐娜。我們和你們有一份進行中的訂單。

妳好，孫小姐。讓我找一下你們的資料。妳可以給我發貨單號碼嗎？

當然。我這裡有。是一〇四九八。

好的。在這裡。你們訂購了一千張毛皮，對嗎？

正確。如果可以的話，我們想要將訂單改成兩千張。

那應該沒問題。你們訂單我們還沒出貨。突然更動是有什麼原因嗎？

如果我們訂多一點的話，似乎可以節省運費。

那總是好的。

確實。我猜想這一點並不會影響價錢吧？或是你可能願意給我們一些折扣？

我很抱歉。我們事實上已經給你們折扣價了。

我瞭解。問問而已不打緊。那麼吐溫先生，感謝你的協助。我很榮幸。

謝謝妳，孫小姐。我們會立即出貨給妳的。

Try it! 換你試試看!

WRITING	請依提示寫出完整句子	☐
LISTENING	請聽 MP3 TRACK 134	☐
SPEAKING	請跟著 MP3 唸唸看	☐

1. Ⓐ You've ordered 2,500 units of halogen light bulbs. Is that correct?

   Ⓑ Yes. ____________________

   (however / make / change / order / OK)

   Ⓐ That shouldn't be a problem. What did you want to change?

2. Ⓐ What was the reason for the change, if you don't mind me asking?

   Ⓑ ____________________

   (clerical error / our part)

3. Ⓐ No problem. I've put in the order change for you.

   Ⓑ ____________________

   (how affect / will / price)

   Ⓐ Double the order will be double the price.

4. Ⓐ Do you think we could order those items as well?

   Ⓑ ____________________

   (you / in luck) (happy / tell you / them / in stock)

答案請參閱第 374 頁

UNIT 27

運送耽擱
Shipment Delay

I. 運送耽擱一定要會的單字片語

MP3 TRACK 135

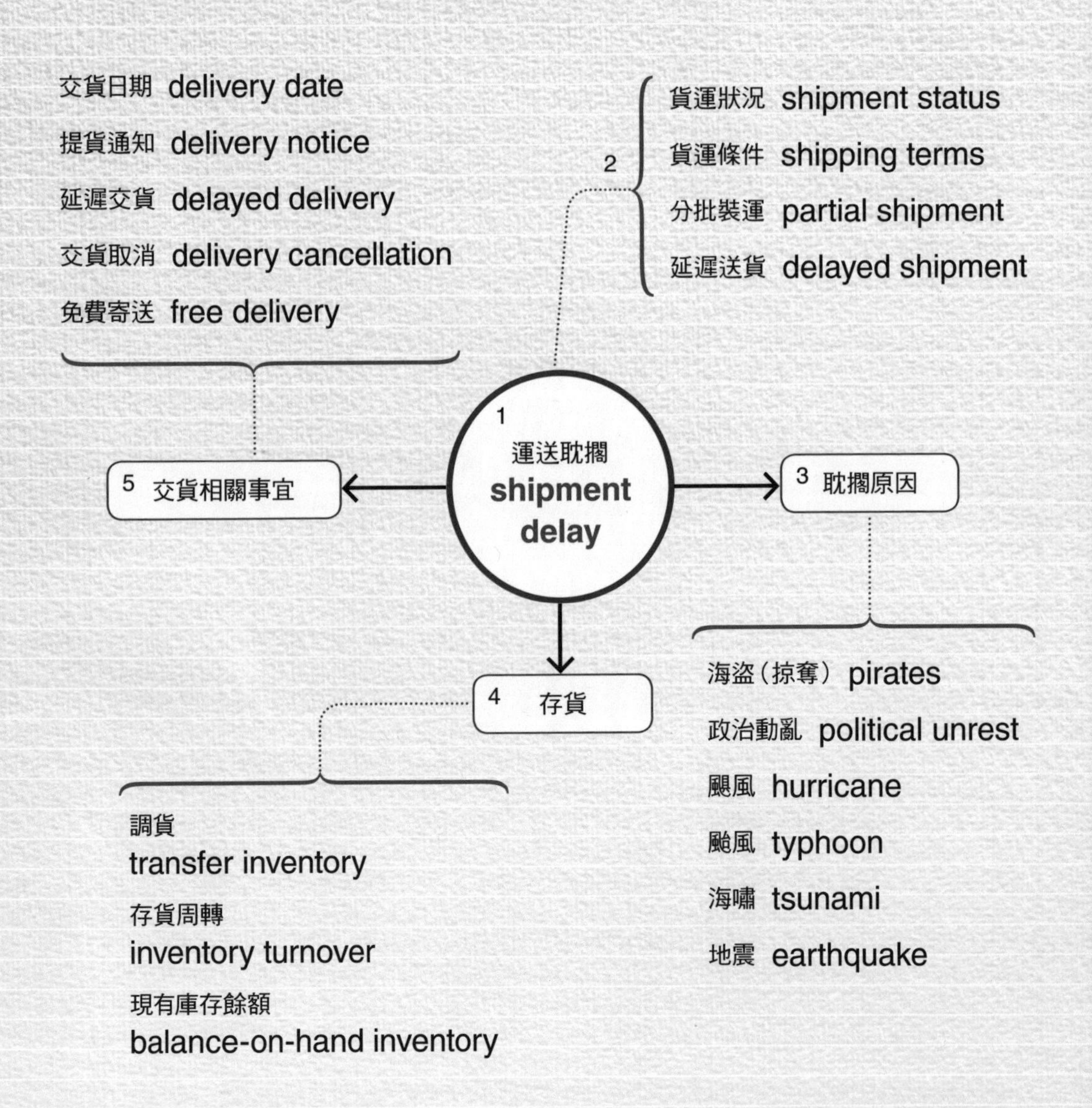

II. 説明運送延誤一定要會的句型

句型1 無法如期履約

I'm sorry to inform you that . . . by + 時間

例 I'm sorry to inform you that we will not be able to complete your order by the date stipulated in the contract.
我很遺憾通知你我們沒有辦法按照合約中所約定的日期來完成你的訂單。

句型2 預估交貨日期

We expect your order can be shipped as early as + 日期

例 We expect your order can be shipped as early as August 14.
我們希望你們的訂單最早可以在八月十四日出貨。

句型3 道歉並提供協助

Please accept my apologies and feel free to contact me . . .

例 Please accept my apologies and feel free to contact me if you have any questions or issues you would like me to address.
請接受我的道歉，如果你有任何疑問或是有任何需要我處理的問題請隨時跟我聯絡。

句型4 希望交貨日期

We do expect to need the shipment by + 時間

例 Based on current sales patterns though, we do expect to need the shipment by early October or so.
但依照目前的銷售模式來看，我們預期十月初左右會需要這批貨。

Ⅲ. 如何用 e-mail 說明運送延誤原因

To: Lance Carrols

Subject: Production/Shipment Delay

Dear Mr. Carrols,

無法如期履約

I'm sorry to inform you that we will not be able to complete your order by the date stipulated[1] in the contract. A serious typhoon recently came through our area of Taiwan. It has caused extensive damage to local infrastructure.[2] In particular, a large landslide has blocked the access road to our factory, making it impossible for workers to get there.

預估交貨日期

We have yet to assess[3] the damage at the factory itself, but we are hopeful that production can continue as soon as the road is cleared. We will update you on developments as soon as there are any. As it stands, we expect your order can be shipped as early as August 14. Be advised, however, this is merely an estimate.

道歉並提供協助

Please accept my apologies and feel free to contact me if you have any questions or issues you would like me to address.

Sincerely,

Michael Meng

Taitung Transistors

Vocabulary & Phrases

1. stipulate [ˋstɪpjəˏlet] *v.* (合約)規定
2. infrastructure [ˋɪnfrəˏstrʌktʃɚ] *n.* 公共設施
3. assess [əˋsɛs] *v.* 評估

中文翻譯

收件人：蘭斯・卡洛斯
主旨：生產／運送延遲

親愛的卡洛斯先生：

我很遺憾通知你我們沒有辦法按照合約中所約定的日期來完成你的訂單。最近有個強烈颱風通過我們台灣這一區，而對本地公共建設造成大規模的破壞。尤其，一場大規模的山崩阻擋了我們往工廠的道路，讓我們的工人無法到達那裡。

我們還沒辦法評估工廠本身的損失，但希望道路一疏通後就能繼續生產。一旦有任何進展我們就會儘快告知你。以目前的狀況來看，我們希望你的訂單最早可以在八月十四日出貨。然而必須告知你，這只是我們的預估。

請接受我的道歉，如果你有任何疑問或是有任何需要我處理的問題請隨時跟我聯絡。

謹上

孟邁克
台東電晶體

文化補給站

根據美國聯邦貿易委員會（U.S. Federal Trade Commission, FTC）消費者保護局（Bureau of Consumer Protection）的法條規定，廠商必須在他們聲明的時間之內送貨。如果沒有承諾時間，也應該要在三十天之內交貨，時間則從接到訂單時開始算。若消費者是使用信用卡付款的話，則廠商交貨日可再寬限二十天（總共是五十天）。

若廠商無法在承諾的時間內送貨，則必須提供消費者「延遲選擇權通知」（delay option notice），此通知應該包含下列要點：

1. 更改後的確切交貨時間，若不確定，則要有無法提供確切交貨日期的聲明。
2. 若廠商無法提供確切交貨時間，需說明下列要點：
   ——延遲的原因
   ——假使消費者選擇接受不確定的延遲，也必須聲明消費者在交貨前有隨時取消訂單的權利
3. 提供消費者取消訂單或全額退費的選擇
4. 消費者取消訂單後的費用由廠商承擔

如果消費者未回應「延遲選擇權通知」，但交貨延遲的時間在三十天之內，則視為接受商品延遲並且願意等候。若延遲已逾三十天，則訂單將被迫取消，並且予以退費。如果廠商發現仍無法在更改後期限內出貨的話，則必須以信件或電話再度通知，並且告知新的交貨日期，或者是取消訂單並給予退費。

Ⅳ. 如何用 e-mail 回覆運送延誤通知信

To: Michael Meng

Subject: Re: Production/Shipment Delay

Dear Mr. Meng,

I am sorry to hear about the difficulties you are experiencing in your country. We at Caltronics wish you the best as you attempt to recover from this natural disaster. The delay in shipment should not cause a problem for us. We are merely filling our inventory. Based on current sales patterns though, we do expect to need the shipment by early October or so. If our order cannot be completed in its entirety by then, perhaps we could arrange a partial shipment. We understand that this must be a difficult situation for you, so we only ask that you keep us informed. Thank you.

希望交貨日期

Again, good luck in your clean-up effort.

Sincerely,

Lance Carrols

Caltronics

中文翻譯

收件人：孟邁克
主旨：回覆：生產／運送延遲

親愛的孟先生：

很遺憾聽到你們國家所遭遇的艱難處境。當你們努力從天然災害中復原時，卡爾電子也祝福你們一切順利。交貨時間的耽擱對我們應該不會造成問題。我們只是要補充庫存。但依照目前的銷售模式來看，我們預期十月初左右會需要這批貨。如果到時候我們的訂單無法全部完成，或許我們可以安排部份出貨。我們理解這對你們而言一定是很困難的處境，所以我們只要求隨時告知我們最新的消息。謝謝你。

再次祝你們的清理工作順利。

謹上

蘭斯．卡洛斯
卡爾電子

Try it! 換你試試看!

1. 很遺憾我們將無法照訂單所開的條件來完成。

2. 我們預估可能會晚大約兩個星期交貨。

3. 若可能造成失望或不便我們深感遺憾。

4. 我們希望我們的訂單最晚可以在五月一日交貨。

答案請參閱第 375 頁

V. 說明運送延誤一定要會這樣說

MP3 TRACK 136

運送問題

A **It's come to my attention that there is a problem with the shipping address we have on file.**

我注意到我們檔案上的寄送地址有問題。

B I think we can help you fix that.

我想我們可以幫你們修正。

預估出貨日期

A **We expect to have your order ready for shipment no later than June 23.**

我們希望可以將你們的訂單在六月二十三日前準備好出貨。

B So can you guarantee delivery by July 10?

所以你們可以保證在七月十日前交貨嗎？

A That should be doable.

應該辦得到。

詢問收貨日期

A **When can we expect delivery?**

預估何時可以交貨？

B That depends on how soon we can get the raw materials for production.

那得看我們能多快拿到生產用的原物料。

希望的交貨日期

A **We need to have those units no later than October 25.**

我們需要在十月二十五日之前收到這些組件。

B We'll see to it that you have them by then.

我們務必會讓你們在那之前收到。

see to 表示「負責；處理」的意思，see to it that S. + V. 則表示「確保……」。

道歉

A **Please accept our sincere apologies for the inconvenience.**

對於造成不便，請接受我們誠摯的道歉。

B We understand. We're sure you'll do your best to get the shipment to us as quickly as possible.

我們瞭解，相信你們會盡全力儘快出貨給我們。

VI. 用電話告知運送延誤

泰森膠帶的維克特打電話通知客戶運送會耽擱，並解釋耽擱原因。

LISTENING 請聽 MP3 TRACK 137 ☐	SPEAKING 請跟著 MP3 唸唸看 ☐

Victor: Hello. Is this Ms. Tsai?

Ali: It is. Who may I ask is calling?

Victor: My name is Victor Danse. I'm in charge of overseeing international orders here at Tyson Tape. You ordered 5,000 rolls of industrial grade tape from us, is that right?

Ali: That is correct. What can I help you with today, Mr. Danse?

無法如期履約

Victor: Well, we may have a problem getting your order to you on time. A key piece of equipment in our factory has broken down and needs to be replaced.

詢問交貨日期

Ali: That's unfortunate. How long of a delay do you expect there to be?

預估交貨日期

Victor: We expect your order to be delayed by a week or two at most.

Ali: I see. We will have to reorganize some of our plans, but it should be feasible.

道歉

Victor: Please accept our apologies for the inconvenience.

Ali: It's OK. You have a great product, and we would be wise to wait for it.

Victor: Thank you for understanding. If you have any questions, just drop me a line or send me an e-mail.

Ali: Thanks for the heads-up. Take care.

妳好，是蔡小姐嗎？

是的，請問你是哪位？

我是維克特・丹斯。我負責監管泰森膠帶的國際訂單。你們向我們訂購了五千捲工業級膠帶，對嗎？

沒錯。丹斯先生，我今天可以幫你什麼忙呢？

是這樣的，我們可能無法準時交貨。我們工廠裡的一個主要設備壞掉了，需要更換。

那太遺憾了。你們預估會延遲多久？

我們預估會晚一個星期、或最多兩個星期才能交貨。

我瞭解了。我們需要將一些計畫重新安排，但應該是可行的。

造成不便請見諒。

沒關係。你們的產品很棒，我們為此等待是明智的。

感謝妳的體諒。如果妳有任何問題，儘管寄短信或寄電子郵件給我。

謝謝你的提醒。保重。

Try it! 換你試試看!

WRITING	請依提示寫出完整句子	☐
LISTENING	請聽 MP3 TRACK 138	☐
SPEAKING	請跟著 MP3 唸唸看	☐

1. Ⓐ We are looking forward to receiving the products.

   Ⓑ ______________________________

   (come to / attention / problem / order)

   Ⓐ What seems to be the issue?

2. Ⓐ When can we expect to receive the shipment?

   Ⓑ ______________________________

   (expect / shipment ready / no later / May 2)

   Ⓐ That should be acceptable, as long as it arrives before the end of that month.

3. Ⓐ This will put us behind our own production schedule. We need those units.

   Ⓑ ______________________________

   (accept / sincerest / apologies)

4. Ⓐ What is the latest delivery date you could accept?

   Ⓑ ______________________________

   (latest / we accept / delivery / April 1)

答案請參閱第 375 頁

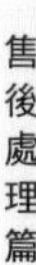

UNIT 28 要求退貨
Return of Defective Goods

I. 要求退貨一定要會的單字片語

MP3 TRACK 139

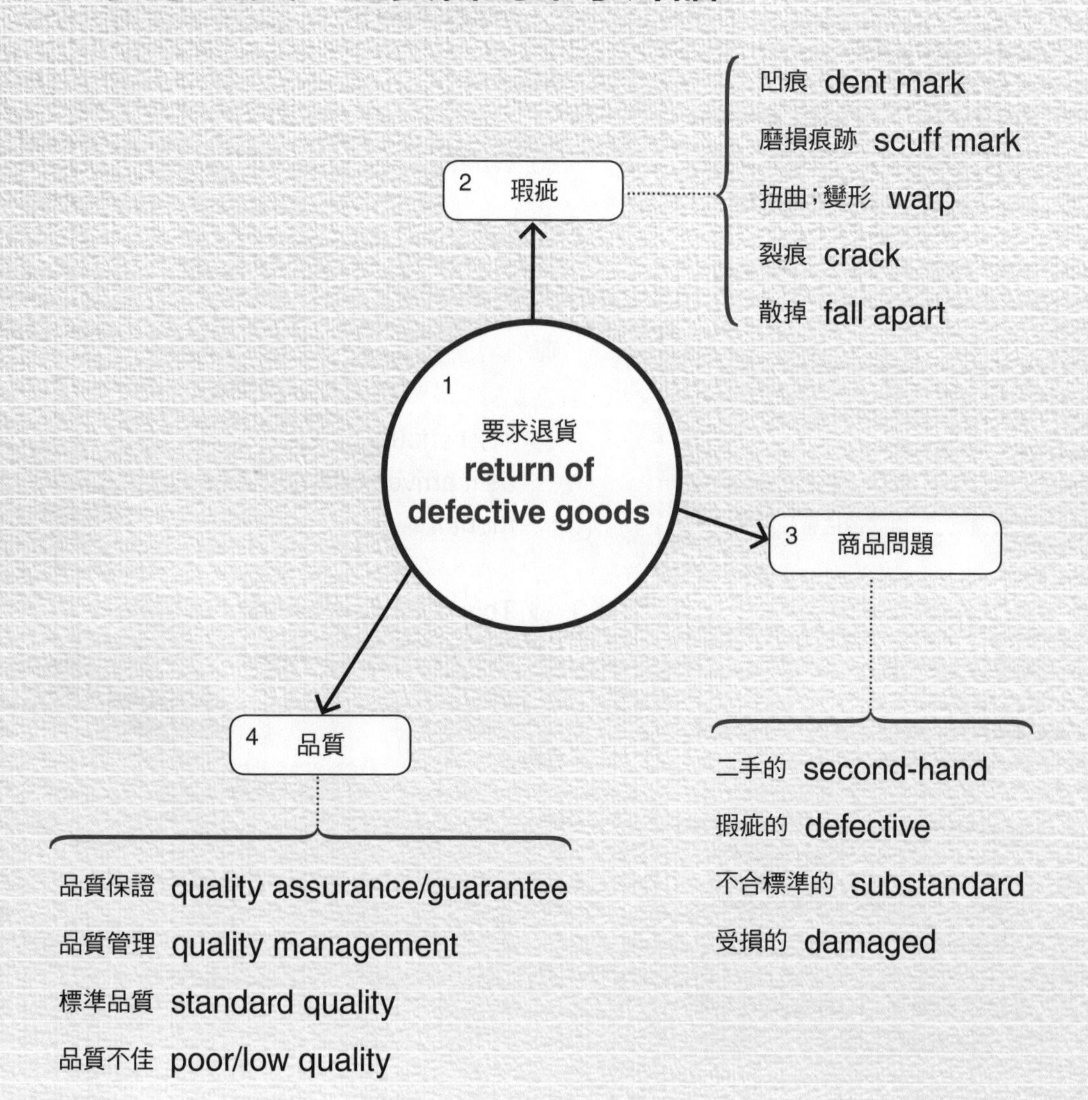

Ⅱ. 要求退貨一定要會的句型

句型1 發現瑕疵品

It has come to our attention that . . .

例 It has come to our attention that your recent shipment of drinking glasses contains defective merchandise.
我們注意到你們最近這批玻璃水杯有瑕疵品。

句型2 說明商品瑕疵狀況

S. **have been found to have** sth（瑕疵狀況）

例 As many as three glasses per crate have been found to have bubbles, cooling marks or sand deposits.
每一箱中就發現多達三個杯子有氣泡、冷卻痕或是沙子沉澱。

句型4 要求退貨

We cordially request a return of sth.

例 We cordially request a return of this shipment.
我們誠摯要求退回這批貨。

句型4 下次交貨日期

Could we expect delivery of the new shipment by + 日期?

例 Could we expect delivery of the new shipment by December 15?
我們可以預計在十二月十五日前收到新貨嗎？

Ⅲ. 如何用 e-mail 要求退貨

To: Greta Kuo

Subject: Defective Merchandise

Dear Ms. Kuo,

發現瑕疵商品

It has come to our attention that your recent shipment of drinking glasses contains defective merchandise. We are extremely disappointed in the quality of work represented therein.[1]

商品瑕疵狀況

As many as three glasses per crate[2] have been found to have bubbles, cooling marks or sand deposits.[3] We consider this amount unacceptable.

要求退貨

As such, we cordially[4] request a return of this shipment.

This order was confirmed on October 6 and received yesterday. We calculate approximately five to six weeks for delivery of a new, non-defective shipment.

下次交貨日期

Could we expect delivery of the new shipment by December 15?

Thank you kindly for your attention to this matter.

Sincerely,

Joseph Holder

Discount Glassware

Vocabulary & Phrases

1. therein [ðɛrˋɪn] *adv.* 在那一點上
2. crate [kret] *n.* 板條箱；箱
3. deposit [dɪˋpɑzət] *n.* 沉澱物
4. cordially [ˋkɔrdʒəli] *adv.* 衷心地；誠懇地

中文翻譯

收件人：郭格蕾塔
主旨：瑕疵商品

親愛的郭小姐：

我們注意到你們最近這批玻璃水杯有瑕疵品，對於商品的品質在這點上我們感到極度失望。每一箱就發現多達三個杯子有氣泡、冷卻痕或是沙子沉澱，我們難以接受這樣的數量。因此我們誠摯要求退回這批貨。

這筆訂單是在十月六日確認並在昨天收到。我們估算一批新的、無瑕疵的貨品約需五至六個星期到貨。我們可以預計在十二月十五日前收到新貨嗎？

衷心感謝妳關切此問題。

謹上

喬瑟夫·侯德
美廉玻璃器皿

文化補給站

guarantee 和 warranty 有何不同？

guarantee 和 warranty 都是保護消費者的機制，在中文中通常也都翻成「保固」，不過兩者有所差別。

guarantee 是廠商提供給消費者的一種法律合約，消費者在購買產品時即享有這項權益，無須另外付錢取得。當消費者發現產品有問題時，可要求廠商退貨或修理，不過 guarantee 通常有一定的期限，但是近來也有廠商推出「終生保固」（lifetime guarantee）。

warranty 則不同，消費者必須額外購買才能享有，通常是在過了保固期之後另外選購的權益，這種保固通常是由零售商或經銷商所提供。

所以兩者最大的差別在於消費者對兩種權益的期待，guarantee 保證產品可以更換，而 warranty 則隱含只提供維修服務。

Ⅳ. 如何用 e-mail 回覆要求退貨信函

To: Joseph Holder

Subject: Re: Defective Merchandise

Dear Mr. Holder,

We are very sorry that you were disappointed in the quality of the glassware you received from us. Our workers do their best to prevent defects at all points of production.

Despite this fact, we are unable to accept your request for a return. First, it is impossible to prevent defects in all units. It would be foolish to discard[1] usable items merely because they are aesthetically[2] imperfect. Second, our quality assurance team inspects all of our shipped products carefully. It is unlikely that they have made many mistakes. Finally, please refer to Section 24B of our contract agreement which states: "Imperfections, including small bubbles, straw marks or cooling marks, are considered acceptable as long as they do not occur in more than 5% of the shipped product." Since there are exactly 100 units in each crate, three per crate is well below the contractual limit.

Absent a breach of contract on our part, we will not be accepting a return on this shipment. If you would like to discuss this issue further, please call me at my office. Thank you very much.

Regards,

Greta Kuo
Taiwan Glass Co.
886-3-3498-6432

你也可以多學一些好用句喔！

Vocabulary & Phrases

1. discard [dɪsˋkɑrd] *v.* 拋棄
2. aesthetically [ɛsˋθɛtɪklɪ] *adv.* 美學觀點地

中文翻譯

收件人：喬瑟夫・侯德
主旨：回覆：瑕疵商品

親愛的侯德先生：

很遺憾你對我們寄去的玻璃杯品質感到失望。我們的工人在每個生產環節上都盡全力避免瑕疵。

儘管如此，我們仍無法接受你的退貨要求。首先，要避免所有的組件沒有瑕疵是不可能的。只因為在美觀上有瑕疵就拋棄這些可用的商品是不明智的。第二，我們的品管團隊都非常仔細檢查我們所有的出貨商品，他們應該不可能出這麼多的錯。最後，請參考我們合約協議上第二十四條B所列：瑕疵商品包括小氣泡、燒痕或冷卻痕的發生率在出貨商品中只要未達百分之五都被視為可接受。由於每箱剛好有一百組，每箱三個則遠低於合約上的限制。

我方並沒有違反合約，我們將無法接受退貨。如果你想要進一步討論這個問題，請打我辦公室的電話。感激不盡。

謹上

郭格蕾塔
台灣玻璃器公司
886-3-3498-6432

Try it! 換你試試看!

1. 我們對於你寄給我們的貨品感到非常失望。

____

____

____

2. 許多組件都發現表面有受損。

____

____

____

(表面 surface)

3. 我們誠摯請求由你們承擔費用，將所有瑕疵商品退回。

____

____

____

4. 我們可以期待最早何時收到新貨呢？

____

____

____

答案請參閱第 375 頁

V. 要求退貨一定要會這樣說

MP3 TRACK 140

抱怨商品品質

A **We are not happy with the quality of the merchandise you sent us.**

對於你們寄來的商品品質我們不是很滿意。

B I'm sorry to hear that. What didn't you like about it?

我很遺憾聽到那樣的事。你們哪一點不滿意？

A **More than a few of the product labels had spelling or grammatical errors.**

不只一些產品標籤上有拼字或文法錯誤。

B Really? I don't see how that is possible.

真的嗎？我看是不太可能的。

詢問哪裡有瑕疵

A **What sort of defects did you find in the products?**

你在商品中發現什麼樣的瑕疵？

B We found misplaced stickers and faded typeface.

我們發現貼紙貼錯位置而且字體褪色。

要求退貨

A **We ask that you grant us a return on this order.**

我們請你們同意退回這筆訂單。

B I'm afraid I'll need more information before I can say yes or no on that.

恐怕我還需要更多訊息才能告訴你可以不可以。

詢問新的交件日期

A **How soon do you expect you could send us our return?**

你預期多快可以將退貨商品寄給我們？

B Assuming we accept your request, in approximately three days.

假使我們接受你們的要求，大概要三天。

VI. 如何用電話要求退貨

帕克配管供應商代表亞當打電話給廠商，表示收到的貨品規格不符，希望能退貨。

LISTENING 請聽 MP3 TRACK 141 ☐ SPEAKING 請跟著 MP3 唸唸看 ☐

Adam: Hello. Is Lydia Hsu there, please?

Lydia: This is she. How may I help you?

Adam: My name is Adam East from Parker Plumbing Supply. We ordered some copper joints from you.

Lydia: Yes, I'm aware of that order. How can I help you Mr. East?

發現瑕疵商品

Adam: Well, there's a problem with the joints you sent us. They're defective.

Lydia: That's impossible. What sort of defect are you referring to?

Adam: They do not fit the pipes we intended to use them with.

Lydia: Perhaps you gave us the wrong specifications.

商品瑕疵狀況

Adam: I don't think so. I reviewed the specifications myself. We would like a return of these joints for correct ones.

Lydia: Before we can accept a return, we'll have to see definitive proof that we've made a mistake.

Adam: In that case, it might be prudent to send a representative here to help us figure out the mistake. What do you think?

Lydia: I'll have to get back to you on that. I don't have the authority to schedule overseas trips. Could I let you know tomorrow?

Adam: That would be fine. Thank you for your time, ma'am.

Lydia: Not at all. Have a good day, sir.

妳好，請問徐莉迪雅在嗎？

我就是，請問有什麼需要幫忙的嗎？

我是從帕克配管供應商打來的，我的名字是亞當· 伊斯特。我們向你們訂購了一些銅製接合器。

是的，我知道那筆訂單。有什麼需要幫忙的嗎，伊斯特先生？

是這樣的，你們送來的接合器有一些問題，它們有瑕疵。

那是不可能的。你指的是哪一種瑕疵？

它們無法安裝在我們想要使用的管子上。

或許你們給我們的規格是錯誤的。

我不這麼認為，我自己檢查過規格。我們想將這些接合器換成正確的。

在我們接受退貨之前，我們必須要看到確切證據證明出錯的是我們。

那麼為了謹慎起見，你們可以派一位代表來幫我們找出錯誤。妳覺得如何？

這事我必須回頭再跟你說。我沒有權力安排海外行程。我可以明天告訴你嗎？

那應該可以。謝謝妳撥空處理，女士。

一點也不會。先生，祝你有個愉快的一天。

Try it! 換你試試看!

WRITING	請依提示寫出完整句子	☐
LISTENING	請聽 MP3 TRACK 142	☐
SPEAKING	請跟著 MP3 唸唸看	☐

1. Ⓐ How may I assist you, Ms. Jenkins?

   Ⓑ ______________________________

   (seems / there / defects / products / sent us)

   Ⓐ What kind of defects are you referring to?

2. Ⓐ What was wrong with the products?

   Ⓑ ______________________________

   (many of / cans / found / dents)

   Ⓐ I find that hard to believe. We pull out dented cans from every shipment.

3. Ⓐ What would you like us to do about it?

   Ⓑ ______________________________

   (we / ask / grant / return / entire shipment)

4. Ⓐ We'd be happy to replace the products for you.

   Ⓑ ______________________________

   (when / expect / receive / new shipment)

   Ⓐ We could probably get it to you as soon as February 22.

答案請參閱第 375 頁

UNIT
29 要求賠償
Request for a Claim

I. 要求賠償一定要會的單字片語

MP3 TRACK 143

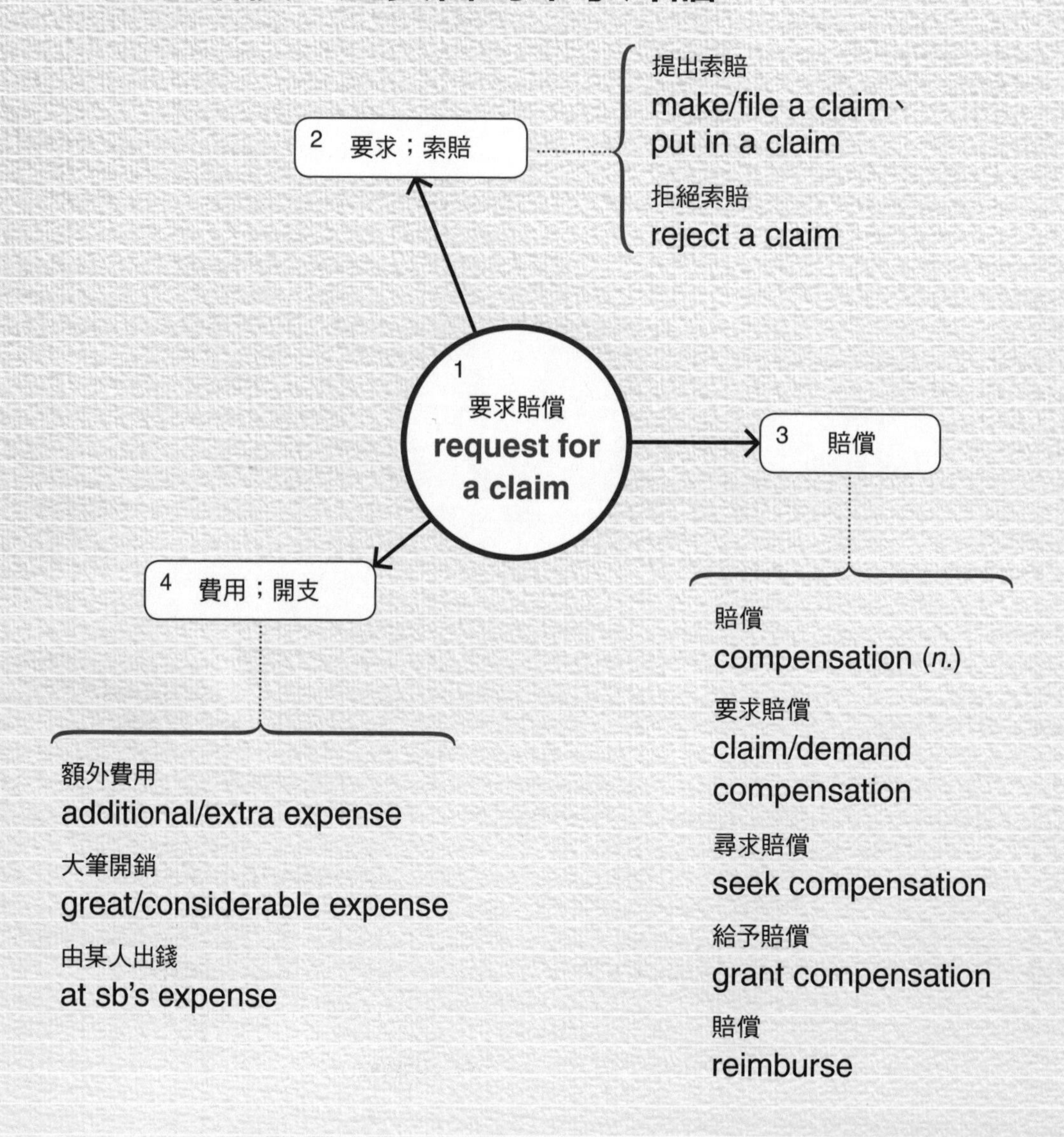

Ⅱ. 要求賠償一定要會的句型

句型 1 要求賠償

. . . **and we are filing a claim** + to V.

例 The machine is malfunctioning, and we are filing a claim to replace it.
機器故障了，我們要求更換。

句型 2 狀況說明

The technician was unable to solve the problem and told us that + S. + V.

例 The technician was unable to solve the problem and told us that the machine may have been improperly assembled.
這位技師無法解決問題並告訴我們機器可能組裝不當。

句型 3 造成損失

As a result of this issue, we are losing sth **and have had to pay for** . . .

例 As a result of this issue, we are losing production time and have had to pay for a technician to come in.
因為這個問題，我們損失了生產時間，還需要付錢請技術人員來。

句型 4 賠償項目

We would like you + to V. **as well as** . . .

例 We would like you to replace the malfunctioning unit as well as reimburse us for the technician's fee and lost production time.
我們想要你們更換故障的組件且賠償我們技師費用及生產時間的損失。

Ⅲ. 如何用 e-mail 要求賠償

To: Harry Rao

Subject: Malfunctioning Machine

Dear Mr. Rao,

This e-mail is to inform you of a problem we've encountered with the T2000 packaging machine we bought from your company. The machine is malfunctioning, and we are filing a claim to replace it. After having it installed by a certified[1] technician, the machine worked fine for a few weeks. Following a storm in which the factory lost power, our personnel found that the error light was on. We called in a professional to fix it. The technician was unable to solve the problem and told us that the machine may have been improperly assembled.[2]

要求賠償

狀況說明

As a result of this issue, we are losing production time and have had to pay for a technician to come in. We need to get a functioning machine installed as quickly as possible. We would like you to replace the malfunctioning unit as well as reimburse us for the technician's fee and lost production time.

造成損失

賠償項目

We hope that you can reply promptly since every day spent is more product we are unable to package and ship. Thank you.

上班族加油站

在抱怨貨物品質時，可用被動的語氣委婉暗示誰該負責，在雙方都心知肚明的情況下盡量避免指名道姓，這是談判技巧之一。

Sincerely,

Erin Satchel

Super Snacks

Vocabulary & Phrases

1. **certified** [ˋsɜtə͵faɪd] *adj.* 認證的
2. **assemble** [əˋsɛmbəl] *v.* 組裝

中文翻譯

收件人：羅哈利
主旨：機器故障

親愛的羅先生：

這封信是要告知你，我們最近向貴公司購買的 T2000 包裝機遇到了問題。機器故障了，我們要求更換。在合格技師組裝過後，機器正常運作了幾個星期。接著在一次暴風雨造成工廠斷電後，我們的同仁發現錯誤燈號亮起。我們打電話找了一位專家來修理。這位技師無法解決問題並告訴我們機器可能組裝不當。

因為這個問題，我們損失了生產時間，還需要付錢請技術人員來。我們需要盡快有一台組裝好能運作的機器。我們想要你們更換故障的組件，並且賠償我們技師費用以及生產時間的損失。

我們希望你們能立即回覆，因為每多花一天的時間，我們就有更多的產品無法包裝運送。謝謝你。

謹上

艾琳・莎翠爾
超級點心

Ⅳ. 如何用 e-mail 回覆索賠信

To: Erin Satchel

Subject: Re: Malfunctioning Machine

Ms. Satchel,

We are sorry to hear that you have been experiencing difficulty with our T2000 packaging machine. It's always ideal to avoid additional expenses whenever possible. Unfortunately, we cannot accept your claim for a return.

It is clear from your description what the problem is. Having lost power and not being shut down properly, the machine required a reboot. If not done precisely, the T2000 may not function. The technician you hired was likely unaware of this. We suggest you refer to the manual supplied with the machine. Pages 115–119 detail the proper procedure for rebooting your T2000. I've attached a copy of the English version of the manual for your convenience. Please have a look.

We understand your frustration. Hopefully, our advice will prove useful. It is our duty and our pleasure to guide you through any technical problems you may have with our machines. We trust that you can resolve the problem with our guidance. If you have any questions or concerns, please do not hesitate to contact me again.

Best Regards,

Harry Rao
Taoyuan Packing Corp.

你也可以多學一些好用句喔！

中文翻譯

收件人：艾琳．莎翠爾
主旨：回覆：機器故障

莎翠爾小姐：

我們很遺憾聽到你們在使用我們 T2000 包裝機時遭遇困難。不管什麼時候，可能的話盡量避免額外的開銷總是最理想的。很遺憾地，我們無法接受你們對於賠償的要求。

從妳的描述中可以很清楚知道問題在哪裡。斷電加上沒有適當關機，機器需要重新開機。如果沒有確切執行，T2000 可能就無法運作。你們所雇用的技師很可能沒有注意到這點。我們建議妳參考機器操作手冊。第一百一十五頁到一百一十九頁詳細說明重新啟動 T2000 的正確程序。我附上了英文手冊讓你們方便查詢。請參閱。

我們理解你們的沮喪。希望我們的建議能有所幫助。引導你們解決使用我們機器所遇到的各種技術問題，是我們職責也是我們的榮幸。相信有了我們的指引你們便能解決問題。如果你們有任何問題或疑慮，請不要客氣，儘管再與我聯繫。

謹上

羅哈利
桃園包裝公司

Try it! 換你試試看!

1. 鼓風機組無法運作，我們需要你修理或是更換。

(鼓風機組 blower unit)

2. 封口機製造不當。

(封口機 sealer)

3. 因為這個問題，我們必須花錢修理。

4. 我們希望你們能對我們的損失提供賠償。

(賠償 compensation)

答案請參閱第 375 頁

V. 要求賠償一定要會這樣說

MP3 TRACK 144

要求賠償

A **The cooling unit has broken down, so we're filing a claim to replace it.**

冷卻裝置損壞了，所以我們要求更換。

B I see. What was the cause of its breaking down?

我瞭解了。是什麼原因造成損壞的？

A We can't tell. It just stopped working one day.

我們不清楚。有一天它就停止運轉了。

說明狀況

A **The product casing seems to have been damaged in production.**

商品外殼好像在製造過程中受損了。

B I don't see how that is possible. It couldn't have made it through inspection.

我看不出有那個可能。那樣不可能通過檢測的。

in production 是指「在生產過程中」，go into production 則表示「投入生產」，即「大量製造」的意思，而 out of production 則表示「停產」，其他說法包括 no longer available、be discontinued。

詢問狀況

A **What kind of difficulty have you experienced with our product?**

你使用我們產品時有遭遇什麼困難嗎？

B There seem to be a number of defects in the way it was made.

似乎在製作的方式上有許多缺失。

蒙受損失

A **Consequently, we've incurred a number of expenses.**

因此，我們蒙受許多的損失。

B Expenses for what?

什麼方面的損失？

incur costs/expenses 表示陷入財務上的困境，incur 的字義是「招致、蒙受」，表示由於某事而導致不好的情況發生。

說明賠償項目

A **We would like you to reimburse us for the cost of repairing the machine.**

我們要你們賠償我們修理機器的費用。

B I understand, but I'm still not sure that the fault lies on our side.

我理解，但我還是不確定錯誤是由我方造成的。

VI. 如何用電話要求索賠

龍之印花的桑妮打電話給貼紙機供應商抱怨機器狀況，希望能拿到修理賠償。

LISTENING 請聽 MP3 TRACK 145 ☐	SPEAKING 請跟著 MP3 唸唸看 ☐

Ron: Hello. This is Ron Reynolds. How may I help you?

Sunny: Hello, Mr. Reynolds. This is Sunny Wu from Dragon Decals. How are you today?

Ron: I'm fine. What can I do for you, Ms. Wu? How is that sticker machine we sent you working out?

狀況說明

Sunny: Actually, that's exactly why I'm calling. It has broken down. Apparently, the cutting blade was dull.

Ron: That's strange. It was a new machine. What would you like us to do about it?

造成損失

賠償項目

Sunny: Well, we've had to pay for a professional to come in and sharpen it. It's working fine now. But we'd like your company to reimburse us for the cost of the sharpening.

Ron: I'm not in a position to authorize that. I'll have to talk to my people.

Sunny: That's fine. You can get back to me once you've had a chance to discuss it with them.

Ron: Thank you for bringing this to our attention.

Sunny: Not at all. It's always a pleasure dealing with your company.

Ron: The feeling is mutual. Until next time, Ms. Wu.

你好。我是榮·雷諾茲。有什麼我可以幫忙的嗎？

你好，雷諾茲先生。我是龍之印花的吳桑妮。你今天好嗎？

我很好。吳小姐，我可以為妳做什麼嗎？我們送去給你們的貼紙機運作如何？

事實上，我正是為此打來。它故障了。很明顯裁切刀是鈍的。

奇怪了。那是新的機器。妳希望我們怎麼做？

嗯，我們已經花錢請了一位專業人員來把它磨利一點。現在運作起來沒問題了。但希望貴公司能賠償我們磨刀的費用。

這件事我無權決定。我必須跟我們其他人談一談。

沒問題。一旦你有機會和他們討論後，你可以再回電給我。

感謝妳告訴我們這件事。

一點也不會。能和你們公司做生意一直都是我的榮幸。

彼此彼此。再聯絡，吳小姐。

Try it! 換你試試看！

WRITING 請依提示寫出完整句子	☐
LISTENING 請聽 MP3 TRACK 146	☐
SPEAKING 請跟著 MP3 唸唸看	☐

1. Ⓐ What seems to be the problem?

 Ⓑ ______________________________

 ____

 (machine / break down / file claim / replace)

 Ⓐ Do you have any idea why it stopped working?

2. Ⓐ Do you have an idea what the problem might be?

 Ⓑ ______________________________

 ____

 (conveyor belt / build / defective materials)

3. Ⓐ Give me a few days so I can talk with my people.

 Ⓑ Please get back to me as soon as possible. ______________________________

 ____

 (lose / at least / a week / production / due to / issue)

4. Ⓐ What would you like us to do about the malfunction?

 Ⓑ ______________________________

 ____

 (we / like / reimburse / cost / repairs)

 Ⓐ I'm not sure I can authorize that just yet.

答案請參閱第 375 頁

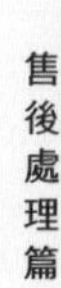

UNIT 30

請求技術協助
Request for Technical Assistance

I. 請求技術協助一定要會的單字片語 MP3 TRACK 147

操作手冊
operating manual/guide

技術手冊
technical handbook

使用者手冊
user's manual/guide

規格指南
specifications manual

故障
breakdown、malfunction (*n.*)

失靈；無法運作
break、die、fail、stop working (*v.*)

故障的
out of order (*adj.*)

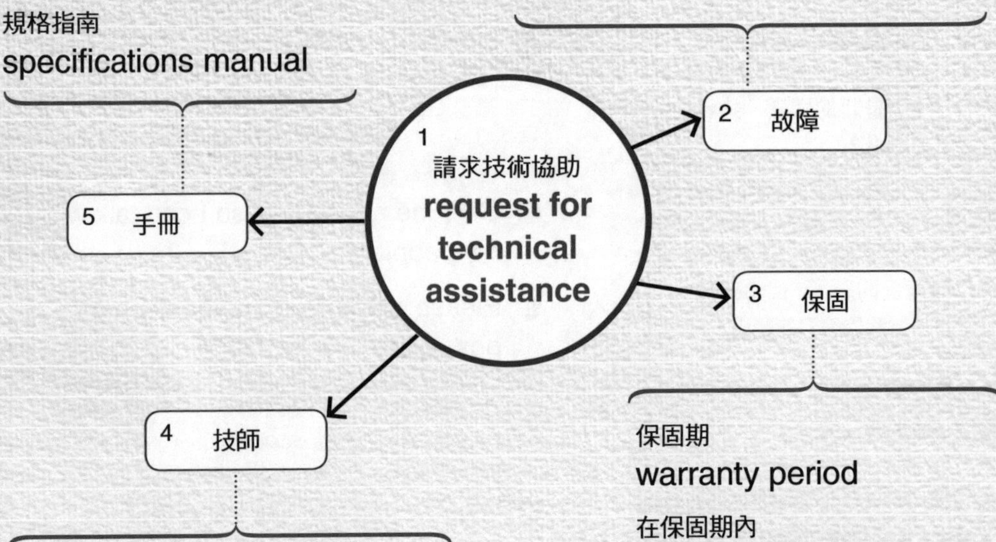

有照技師 licensed technician

無照技師 unlicensed technician

授權技師 authorized technician

本地技師 local technician

保固期
warranty period

在保固期內
under warranty

不在保固期內
no longer under warranty

終身保固
lifetime warranty

Ⅱ. 請求技術協助一定要會的句型

句型 1　遇到問題

We are currently experiencing a problem with sth **and require** . . .

例 We are currently experiencing a problem with the mixer and require technical assistance.
我們最近在攪拌器使用上遇到問題，需要技術協助。

句型 2　聯絡窗口

Our order was handled by sb.

例 Our order was handled by your sales representative, Mr. Terrence Powder.
我們的訂單是由你們的業務代表泰倫斯．鮑德所負責。

句型 3　推測原因

Our best guess is that there is an internal problem with sth.

例 Our best guess is that there is an internal problem with the motor that drives the mixer.
我們認為最有可能是啟動攪拌棒的馬達裡頭有問題。

句型 4　引用條款

S.（保證書）**states:**

例 The product warranty states:
商品保證書上寫著：

Ⅲ. 如何用 e-mail 請求技術協助

To: Great Grinders

Subject: Requesting Technical Assistance

Greetings,

Our company purchased a meat grinder from you about eight months ago. We are currently experiencing a problem with the mixer and require technical assistance. We also have a few questions regarding how to proceed. Our order was handled by your sales representative, Mr. Terrence Powder, but we have been unable to reach him by phone or e-mail. Perhaps he could fill you in on the details of our order.

遇到問題

聯絡窗口

Back to the problem at hand: Not long ago, the grinder's mixing paddles started slowing down. Then, according to the operator at the time, the machine made a strange noise, and it stopped mixing. The other components seem to be fully functional. Our best guess is that there is an internal problem with the motor that drives the mixer.

推測原因

The product warranty states: *"In the event of a malfunction or other complication with the machinery, the client may request technical service within a year of purchase. This may take the form of instructions, advice or guidance. If the matter requires the attention of a technician, support may be sent from Great Grinders directly or outsourced to a local company."* Given our location, we assume you would prefer that we find a certified technician here and then ask you to reimburse us. Do you know of any in our area, or could you inform us of how to find one?

引用條款

Finally, we ask that you give us an answer no later than Friday, May 4, if possible. We are currently using our back-up grinder which is considerably slower than yours. This has made it difficult to meet production demands. We look forward to hearing back from you shortly.

Sincerely,

Aaron Hsieh
Magic Meats

中文翻譯

收件人：優良絞碎機
主旨：請求技術協助

您好：

我們公司在大約八個月前向你們購買了絞肉機。我們最近在攪拌器使用上遇到問題，需要技術協助，關於如何進行此事我們也有一些疑問。我們的訂單是由你們的業務代表泰倫斯·鮑德所負責，但是我們一直無法透過電話或是電子郵件聯絡上他，或許他可以詳盡提供你一些我們訂單上的細節。

回到手邊的問題：不久之前，絞肉機的攪拌棒開始速度變慢。然後，根據那時候的作業員所說，機器發出奇怪的聲音，之後就停止攪拌，其他組件的功能似乎都正常，我們認為最有可能是啟動攪拌棒的馬達裡頭有問題。

商品保證書上寫著：在購買一年內，如有故障或是其他機器引發的問題，客戶可以要求技術服務。服務方式包括操作說明、建議或是諮詢。如果問題需要由技術人員處理，可以從優良絞碎機直接派人協助或是外包給當地的公司。以我們的地點來說，我們猜想你們寧願選擇我們在這裡找到合格的技術人員，然後再向你們求償。你知道在我們這區有沒有這樣的人員，或是能不能告訴我們要如何找到這樣的人？

最後，如果可能的話，我們請求你們在五月四日星期五之前給我們一個答覆。我們目前使用的是備用絞肉機，和你們的機器比起來慢多了。這使我們難以達到生產需求。我們期待盡快聽到你的回覆。

謹上

謝艾倫
神奇肉品

Ⅳ. 用 e-mail 回覆請求技術協助信

To: Aaron Hsieh

Subject: Re: Requesting Technical Assistance

Dear Mr. Hsieh,

Thank you for contacting us concerning your problem with the meat grinder we sent you.

Mr. Powder is no longer with Great Grinders, so I will be handling our correspondence from now on. We do have your invoice on file, and you will be pleased to know that the warranty does indeed cover technical assistance in this case.

Finding a technician in your area should not be difficult. The engineering behind our grinders is quite standard in the industry. Any technician certified to service these kinds of machines should be more than capable of handling your problem. We do ask, however, that you get an estimate before going ahead with any repairs.

I understand that you want this issue resolved quickly, so please call me as soon as you know the price of the repairs. That way, I can verbally authorize it for you on the spot. You can find my number below. I will be happy to answer your call at any time of day. Service is our number one priority.

Thank you.

Regards,

Kelly Waite
Great Grinders

你也可以多學一些好用句喔！

中文翻譯

收件人：謝艾倫
主旨：回覆：請求技術協助

親愛的謝先生：

關於我們寄去絞肉機的問題，感謝你與我們聯繫。

鮑德先生已經不在優良絞碎機服務，所以今後將由我負責處理我們之間的聯繫。我們檔案中有你們的發貨單，而你會很高興知道，保固確實包含了此種情況下的技術協助。

要在你們地區找到技術人員應該不是件難事。我們絞碎機的工程技術是相當符合工業標準的。任何一位提供此種機器服務的合格技術員應該都能勝任處理你的問題。不過，我們必須要求你在進行任何維修前要先拿到估價。

我瞭解你們想盡快解決這個問題，所以請你一知道維修費用就盡快打給我。那樣的話我可以當下口頭允諾你。你可以在信件下方找到我的號碼。一天之中我隨時都很樂意接到你的電話。服務是我們的最高指導原則。

謝謝

謹上

凱莉· 維特
優良絞碎機

Try it! 換你試試看!

1. 我們在操作機器上遭遇困難。

2. Rick Tally 是你們公司和我們生意往來的人。

(生意往來 deal with)

3. 產品手冊說每天晚上要將機器關機。

4. 似乎是和程式有關的問題。

答案請參閱第 376 頁

V. 請求技術協助一定要會這樣說

MP3 TRACK 148

遇到問題

A **We've been having some problems with the labeler you sent us.**

你寄來的標籤機我們有一些問題。

B That's not good. Can you describe the problem in more detail?

那不太妙。你可以更詳細說明是什麼問題嗎？

聯絡窗口

A **We have previously been in contact with Ms. Tina Wu, your Head of Sales.**

我們先前一直是和吳蒂娜，你們銷售部的經理聯繫。

B Yes, Ms. Wu. I can refer your issue to her if you'd like.

是的，吳小姐。如果妳要的話，我可以把妳的問題轉給她。

refer . . . to sb 表示將某事提交給有權處理的人。

引用條款

A **By the terms of our agreement, we are entitled to free technical assistance.**

根據合約條款，我們享有免費技術協助的權利。

B That's true. We'll do our best to help you handle the situation.

沒錯。我們會盡我們所能來幫你處理問題。

推測原因

A **We suspect the problem has something to do with the circuitry.**

我們懷疑問題跟電路有關。

B That could very well be.

那是非常有可能的。

回覆期限

A **We hope you can give us an answer now or at least by the end of the day.**

我們希望你現在或至少在今天之前可以給我們一個答案。

B I can let you know by the end of the workday.

在今天下班前我會讓你知道。

A Your time or ours?

你們的時間還是我們的時間？

B Whichever you prefer.

看你比較希望是哪一個。

「工作日」還可以用 business day、working day 來表示。

VI. 如何用電話請求技術協助

冰淇淋小屋代表葛瑞格打電話給供應商，詢問機器維修問題是否包括在保證條款中。

LISTENING 請聽 MP3 TRACK 149 ☐ SPEAKING 請跟著 MP3 唸唸看 ☐

Greg: Hello, this is Greg Lin from the Ice Cream Hut. We bought one of your machines a few months ago.

Wanda: That's right. Hi, Mr. Lin. What can I do for you today?

Greg: Please. Call me Greg. Actually, we've got a problem with the machine you sent us.

遇到問題

Wanda: That's not good. What kind of problem exactly?

Greg: It's an issue with the dispenser. It doesn't release the ice cream evenly. We haven't been able to fix it ourselves.

推測原因

Wanda: Alright. What would you like to see from our end?

Greg: We understand that by the terms of our warranty we're entitled to free technical assistance.

引用條款

Wanda: That is correct. We can't send our own technician over due to the distance, but we can reimburse you for one outsourced from your area.

Greg: Great. That's exactly what I wanted to hear.

Wanda: Do me a favor though and call me while the technician is there so we can maybe suggest ways to keep the cost down.

Greg: Excellent. Thanks a lot.

Wanda: OK. Just call me whenever you're ready.

妳好。這裡是冰淇淋小屋的林葛瑞格。幾個月前我們買了你們的一部機器。

沒錯。林先生你好。今天我可以為你做什麼嗎？

請叫我葛瑞格就好。事實上，你們寄來的機器我們有一些問題。

那不太妙。確切來說是什麼問題呢？

是分配器的問題。它無法將冰淇淋均勻擠壓出來。我們沒有辦法自己修理。

好的。你想要我們怎麼做？

我們知道依據保固條款，我們有權利享有免費的技術協助。

沒錯。由於距離的關係我們沒有辦法派自己的技術人員過去，但可以補償你們所在地區一位外包人員的費用。

太好了，這就是我想要聽到的。

但請幫我一個忙，當技術人員到你那邊的時候請打給我。我們或許可以給一些建議好降低成本。

太棒了。感激不盡。

好的。任何時候你準備好就打給我。

Try it! 換你試試看!

WRITING 請依提示寫出完整句子	☐
LISTENING 請聽 MP3 TRACK 150	☐
SPEAKING 請跟著 MP3 唸唸看	☐

1. Ⓐ What did you want to discuss?

   Ⓑ ______________________________

   (we / problems / sander / buy from you)

   Ⓐ I'm sorry to hear that. What kind of problems?

2. Ⓐ Could you tell me the name of the representative who assisted you with your purchase?

   Ⓑ ______________________________

   (our order / handle / sales representative / Ms. Tammy Cincher)

   Ⓐ Ms. Cincher is not with us anymore. I'd be more than happy to help you though.

3. Ⓐ What would you like to see from our end?

   Ⓑ ______________________________

   (product warranty / mention / right / technical assistance)

4. Ⓐ What do you think might be the problem?

   Ⓑ ______________________________

   (we / suspect / issue / wiring)

5. Ⓐ I think I can send you one of our technicians to look at the problem. I'll let you know.

   Ⓑ ______________________________

   (we / hope / let know / right away / as soon as possible)

答案請參閱第 376 頁

解答頁 Answer Key

人際關係篇

Unit 1 請假通知

✉ 換你試試看 —p. 43

1. I want to confirm that I will be out of the office starting on May 1.
2. I am currently working on a review of the sales department's quarterly budget.
3. My colleague, Joe Hu, will be stepping in to handle my responsibilities while I am away.
4. In case of an emergency, you can contact me at the following number: (312) 339-4558.

換你試試看 —p. 47

1. I will be out of the office for one week starting next Monday.
2. I am currently working on an analysis of our overseas branches.
3. My colleague, Doris Zhou, will be filling in for me while I'm away.
4. In case of an emergency, you may contact me at my mother's house at (312) 443-6786.

Unit 2 表達祝賀

✉ 換你試試看 —p. 55

1. I wanted to congratulate you on your recent marriage.
2. Congratulations to you on your recent award.
3. I was pleased to hear this good news.
4. I wish you continued success in the future.

換你試試看 —p. 59

1. Congratulations on the big sale.
2. Congratulations on your recent wedding.
3. That's great to hear. Congratulations on becoming a new father.
4. You deserve it! Congratulations on this honor.

Unit 3 表達感謝

✉ 換你試試看 —p. 67

1. I am writing to express my gratitude for your help.
2. Thank you for your thoughtful gift.
3. If I can ever assist you in any way, please don't hesitate to let me know.
4. I look forward to returning the favor in the future.

換你試試看 —p. 71

1. Yes, and I want to express my gratitude for your help.
2. Thank you for your thoughtful gift.
3. If I can ever assist you in any way, please don't hesitate to let me know.
4. Yes, and I look forward to returning the favor in the future.

Unit 4 表達慰問

✉ 換你試試看 —p. 79

1. I can't tell you how sad I am to learn about your recent tragedy.
2. I hope you are feeling better now and recovering quickly.
3. Please know that you have a friend you can rely on if you need help in any way.
4. Our thoughts are with you during this difficult moment.

換你試試看 —p. 83

1. Yes, I just saw the news about the earthquake on TV.
2. I was sorry to hear about your recent health issue.
3. Please remember that I am always here for you.
4. Our prayers are with you.

Unit 5 弔唁

✉ 換你試試看 —p. 89

1. We were so sorry to hear about your great loss.
2. Please accept my sincere condolences.
3. John will be sadly missed by everyone who knew him.
4. We are remembering you and your family in our prayers.

換你試試看 —p. 93

1. Yes, I was saddened to learn of the loss in his family.
2. Please allow me to express my deepest sympathy to you at this time.
3. I was sorry to hear about his passing.
4. I am here for you if you need anything.

Unit 6 邀請

✉ 換你試試看 —p. 101

1. I would like to invite you and your wife to our Thanksgiving dinner party.
2. The wedding ceremony will be held on June 15, 2012, at the Royal Regency Hotel.
3. There is no need to bring a gift.
4. We hope you can join us on this special day.

換你試試看 —p. 105

1. I'm calling to invite you to our company's Christmas party.
2. The party will be held at my apartment on Friday at 7 p.m.
3. The dress code for the wedding is formal.
4. I hope to see you there.

商務往來篇

Unit 7 展開業務關係

換你試試看 —p. 113

1. We provide a wide range of high-quality stationery products to Hong Kong, Singapore, Taiwan, and the Philippines.
2. Our sales volume exceeded eight million dollars last quarter.
3. You should know that our company won the Best Asian Business Award last year.
4. We would like to welcome you to become a part of our fast-growing business.

換你試試看 —p. 117

1. We produce solar panels for the industrial sector.
2. Our sales volume passed 400,000 units last year.
3. You should know that we won the Marshall Award three years in a row.
4. We would like to invite you to attend our product demonstration.

Unit 8 確認業務拜訪

換你試試看 —p. 123

1. Just a reminder that I will be visiting your company to introduce our newest products.
2. I confirm that our meeting is at 10 a.m. on Friday, December 30, at your office.
3. Would it be possible to request an overhead projector for my presentation?
4. I will get in touch with you again next week.

換你試試看 —p. 127

1. I am calling to remind you of my upcoming visit to your company to introduce our company's latest products.
2. To confirm, I am scheduled to meet you next Friday at 2 p.m. at your office.
3. Yes, I'd like to know if there's a projection screen in the conference room.
4. I will contact you again on Friday to reconfirm everything.

Unit 9 邀請蒞臨貿易展

換你試試看 —p. 133

1. I am writing to invite you to visit us at our booth at the upcoming Taipei Trade Show next month.
2. We will have our new line of products on display at this trade show.
3. Our booth number is 125, and it will be located in aisle B, Trade Hall 3.
4. Don't forget to bring along the attached coupon for a chance to win a big prize.

换你試試看 —p. 137

1. I would like to invite you to visit our stand at the upcoming International Machinery Tool Show next month.
2. We will be launching exciting new products and offering hands-on demonstrations.
3. You can find our display in booth 25, located in Trade Hall 1.
4. If you have any questions, please feel free to contact me or anyone in our marketing department.

Unit 10 安排會面

換你試試看 —p. 143

1. I have arranged for you to stay at the Belmont Suites Hotel for the duration of your visit.
2. I have also planned a schedule for your three-day visit.
3. I want to remind you that the weather is very warm now in Taipei, so you may want to pack lightly for your trip.
4. I hope you have a pleasant flight.

換你試試看 —p. 147

1. I have already reserved a suite for you at the Royal Inn for the length of your stay.
2. I have come up with an itinerary for your visit.
3. It can get quite chilly and rainy in February.
4. I wish you a smooth journey.

Unit 11 取消會面

換你試試看 —p. 153

1. I am sorry, but I must cancel my trip next month to Hong Kong.
2. I have an emergency in the family that I must attend to.
3. I would like to postpone my travel plans to a later date, March perhaps.
4. If you'd like, we could set up a video conference.

換你試試看 —p. 157

1. I am very sorry, but I am calling to cancel my upcoming visit to Taiwan.
2. Unfortunately, an urgent matter has come up here in my company, and I must deal with it.
3. Could we possibly reschedule for October?
4. If you would still like to discuss the matters we had scheduled, perhaps we can arrange for a teleconference via Skype.

Unit 12 感謝招待

換你試試看 —p. 163

1. I wanted to thank you for making my trip to Taiwan a success.
2. It was so nice of you to take us around the city.
3. We especially enjoyed seeing Taipei 101.
4. I look forward to reciprocating your hospitality the next time you're in Chicago.

換你試試看 —p. 167

1. I am calling to let you know what a wonderful trip I had.
2. I really appreciate all of your help in making my trip such a success.
3. It was great to meet with you in person to discuss our business.
4. Please remember that you are always welcome as our guests.

Unit 13 請求代理權

換你試試看 —p. 173

1. I am writing to express my interest in distributing and selling your company's products in Taiwan.
2. We, at TLC Distributors, have demonstrated consistent growth over the last eight years, with five branch offices now across Southeast Asia.
3. We will not only increase your overall sales volume, but we will bring you many other advantages that only we can provide.
4. To further discuss this opportunity for growth, please don't hesitate to contact me.

換你試試看 —p. 177

1. I am calling to discuss the possibility of acquiring exclusive agency for your company.
2. We have been in business for over seven years.
3. With us as your exclusive agent, you can enjoy the benefits of our thorough knowledge of the local market.
4. Exclusive agency certainly offers many advantages for your company.

Unit 14 請求著作權許可

換你試試看 —p. 183

1. I am writing to request permission to use the following photos in my article.
2. We would be willing to pay a small licensing fee for permission to reprint the article.
3. Please let me know how you would like us to attribute your work.
4. If you are not the copyright owner, I would appreciate any information you can give me about the proper person or company to contact.

換你試試看 —p. 187

1. I am calling to request permission to reproduce excerpts of your book on our Web site.
2. Please let us know if we should pay a flat fee or royalties for use.
3. Of course, we will be sure to attribute the work correctly.
4. Are you also the copyright owner of these articles?

Unit 15 介紹新的業務關係

✉ 換你試試看 —p. 193

1. We want to inform you of the new partnership between Peter Barnes Consultancy and Domino Corporation.
2. The two companies plan to merge at the end of 2012.
3. By combining both companies, we will be able to give our customers more benefits.
4. I will be moving to a new sales team, but I will remain as your primary contact.

換你試試看 —p. 197

1. We would like to inform you of our upcoming merger with Venture Intelligence Company.
2. The newly merged company will be named Apex Capital Ventures.
3. This merger will make the company stronger and more competitive on the market.
4. Your status as our valued partner will remain unchanged.

貿易活動篇

Unit 16 詢問產品

✉ 換你試試看 —p. 205

1. I received your contact information from Judy Jenkins, whom I believe you know well.
2. I'm particularly impressed with your Art-Deco Wall Stickers.
3. My only concern is product safety. We need them to conform to CPSC standards.
4. Could you provide me with the pricing information?

換你試試看 —p. 209

1. I received your information from Cathy Fletcher.
2. I am particularly interested in the Noland swivel fans.
3. My only concern is size. We need extremely small units.
4. Could you also send me the pricing information for those products?

Unit 17 詢問價格

✉ 換你試試看 —p. 215

1. We're considering placing an order for a few of your products.
2. Could we negotiate a discount if we place a large order?
3. What assurance can you give us against defective products?
4. How would you deal with defective merchandise?

換你試試看 —p. 219

1. We're interested in placing an order for the HearWell headphones.
2. If we ordered one thousand units, could you give us a discount?
3. Yes. I'd like to know if you provide a guarantee against defects?
4. But we need to know precisely when the order will be shipped.

Unit 18 報價後續追蹤

✉ 換你試試看 —p. 225

1. We're still waiting to hear your final decision on this order.
2. You won't find a better or more reliable battery from anyone else.
3. We hope to receive a reply from you by next week.
4. Our offices have been closed due to a typhoon.

換你試試看 —p. 230

1. We haven't received a confirmation from you yet.
2. You won't find a softer tissue on the market.
3. We'll be happy to assist you in any way possible.
4. Our inventory manager has been out sick for the past three days.

Unit 19 議價

✉ 換你試試看 —p. 237

1. Unfortunately, our funds are limited for this order.
2. We need to lower our order to 15,000 units.
3. Would it be possible to offer a discount?
4. We would like to come to a deal quickly.

換你試試看 —p. 241

1. The quoted price is a little out of / over our budget.
2. Without a rate adjustment, we will need to lower our order.
3. Is there any way you could lower the price a bit?
4. We would like to come to a deal as soon as possible.

Unit 20 索取樣品

✉ 換你試試看 —p. 247

1. This letter is to request samples of your blue paint colors.
2. We are solely/only interested in the metal models.
3. Please send the samples to this address: No. 201, Sec. 6, Zhongxiao East Road, Taipei.
4. Do you need to reimburse for the cost of the samples?

換你試試看 —p. 252

1. I'm calling to request samples of your novelty clocks.
2. We're not interested in the miniature models.
3. Please send the samples to this address. Do you need to get a pen?
4. How would you like us to pay for the samples?

Unit 21 通知價格調漲

✉ 換你試試看 —p. 259

1. Please accept this letter as notification of an impending rate adjustment.
2. The increase is due to higher costs of raw materials.
3. There is no better source for long-lasting batteries.
4. Why not take this opportunity to place an order before the price increase takes effect?

換你試試看 —p. 264

1. I'm sorry to inform you of a slight rate adjustment effective May 21.
2. It's the result of higher shipping costs.
3. Our footwear is the best on the market.
4. This is your last chance to get our products at the same low prices.

Unit 22 詢問其他付款方式

✉ 換你試試看 —p. 271

1. I'm not exactly sure how to explain our situation.
2. For larger shipments, we prefer to receive payment in advance.
3. We look forward to hearing your reply.
4. We are afraid we don't have the funds to make a full payment in advance.

換你試試看 —p. 275

1. I'm not sure how to explain the situation.
2. For this order, we'd like to receive payment in advance.
3. We hope you can give us an answer soon.
4. We are concerned about additional shipping costs.

Unit 23 下訂單

✉ 換你試試看 —p. 281

1. I would like to place an order for your 16-ounce Wood Handle Hammers.
2. We'd like to start with an order of 100 pieces.
3. Could you lower the price if we ordered 150 pieces?
4. When would it be possible to receive the order? /
   When can I expect to receive the order?

換你試試看 —p. 285

1. I'd like to place an order for the A-51 office chairs.
2. We wanted to place an order for 30 office chairs.
3. Would there be a discount if we ordered fifty units?
4. Could you tell us when is the earliest we can receive this order?

Unit 24 安排出貨

換你試試看 —p. 289

1. We have a few concerns about the shipping details.
2. Could you provide us with an estimated delivery date?
3. Will we be getting a shipping confirmation?
4. Would it be too much to request gift wrapping?

換你試試看 —p. 296

1. I wanted to discuss the shipping arrangements.
2. What is the expected delivery date for our order?
3. Will we be getting a shipping confirmation?
4. Would it be possible to ship the units individually?

Unit 25 催款

換你試試看 —p. 303

1. Payment for your order is now a week overdue.
2. Your payment of NT$3,563 was due on August 7.
3. If you explain the problem to us, we will be happy to help you resolve the issue.
4. We appreciate your prompt reply.

換你試試看 —p. 307

1. Were you aware that your payment is now overdue?
2. Your total of NT$7,800 was due on December 21.
3. We would greatly appreciate an explanation for the delay in payment.
4. We await your prompt attention to this matter.

售後處理篇

Unit 26 更改訂單

換你試試看 —p. 315

1. Would it be possible to decrease our order by 300 units?
2. This is the result of an accounting error.
3. Please tell us the new price as soon as you've calculated it.
4. It's my pleasure to inform you that your order now qualifies for a discount.

換你試試看 —p. 320

1. However, we'd like to make a change to the order if that's OK.
2. It is due to a clerical error on our part.
3. How will this affect the price?
4. You're in luck! I'm happy to tell you we have them in stock.

Unit 27 運送耽擱

✉ 換你試試看 —p. 327

1. Unfortunately, we will not be able to complete your order as placed.
2. We expect delivery to be delayed by about two weeks.
3. We deeply regret any disappointment or inconvenience this may cause.
4. We hope to have our order delivered by May 1 at the latest.

換你試試看 —p. 331

1. It has come to my attention that there is a problem with your order.
2. We expect to have your shipment ready no later than May 2.
3. Please accept our sincerest apologies.
4. The latest we can accept delivery is April 1.

Unit 28 要求退貨

✉ 換你試試看 —p. 337

1. We are extremely disappointed in the shipment you sent us.
2. A number of units were found to have damage on the surface.
3. We kindly/cordially request a return of all defective merchandise at your expense.
4. When would be the earliest we could expect delivery of the new shipment?

換你試試看 —p. 342

1. It seems there are defects in the products you sent us.
2. Many of the cans were found to have dents.
3. We ask that you grant us a return of the entire shipment.
4. When could we expect to receive the new shipment?

Unit 29 要求賠償

✉ 換你試試看 —p. 349

1. The blower unit doesn't work, and we need you to fix it or replace it.
2. The sealer was improperly built/manufactured.
3. Because of this problem, we've had to pay for repairs.
4. We hope you can provide compensation for our losses.

換你試試看 —p. 353

1. The machine has broken down and we are filing a claim to replace it.
2. The conveyor belt was built with defective materials.
3. We've lost at least a week of production due to this issue.
4. We would like you to reimburse us for the cost of the repairs.

Unit 30 請求技術協助

換你試試看 —p. 359

1. We are experiencing difficulty operating the machine.
2. Rick Tally was the person we dealt with at your company.
3. The product manual says to turn the machine off every night.
4. It seems to be an issue with the programming.

換你試試看 —p. 364

1. We've been having problems with the sander we bought from you.
2. Our order was handled by your sales representative Ms. Tammy Cincher.
3. The product warranty mentions our right to technical assistance.
4. We suspect it's an issue with the wiring.
5. We hope you can let us know right away or as soon as possible.

國家圖書館出版品預行編目 (CIP) 資料

上班族不能不會的 E-mail + 電話英語 /
Timothy Daniel Ostrander, Ted Pigott 作 ;
王琳詔總編輯 . —— 初版 . ——
臺北市 : 希伯崙公司 , 民 100.12

面 ;　公分 .

ISBN　978-986-6051-17-3（平裝附光碟片）

1. 商業書信 2. 商業英文 3. 商業應用文 4. 電子郵件

493.6　　100024874

《上班族不能不會的 E-mail + 電話英語》讀者回函卡

謝謝您購買 LiveABC 互動英語系列產品

如果您願意，請您詳細填寫下列資料，免貼郵票寄回LiveABC即可獲贈《CNN互動英語》、《Live互動英語》、《每日一句週報》電子學習報3個月期（價值：900元）及LiveABC不定期提供的最新出版資訊。

姓名		性別	□男 □女
出生日期	年 月 日	聯絡電話	
住址	□□□		
E-mail			
學歷	□國中以下 □國中 □高中 □大專及大學 □研究所		
職業	□學生 □資訊業 □工 □商 □服務業 □軍警公教 □自由業及專業 □其他＿＿＿＿		

您從何處得知本書？
□書店 □網站 □電子型錄 □他人推薦 □雜誌 □其他＿＿＿＿

您以何種方式購得此書？
□一般書店 □連鎖書店 □網路 □郵局劃撥 □其他＿＿＿＿

您覺得本書的價格？ □偏低 □合理 □偏高

您對本書的評價

	書名	封面	內容	編排	紙張
很滿意	□	□	□	□	□
還不錯	□	□	□	□	□
普　通	□	□	□	□	□
不滿意	□	□	□	□	□
很後悔	□	□	□	□	□

您希望我們製作哪些學習主題？

您對我們的建議：

黏　貼　處

廣　告　回　信
台北郵局登記證
台北廣字 1194 號

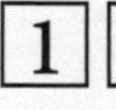

105

台北市松山區八德路三段32號12樓

希伯崙股份有限公司客戶服務部　收

縣市　市區鄉鎮　村里路街　段　鄰巷　弄　號　樓　室